The Countryman
Wild Life Book

Cover Photograph by Jane Burton/Bruce Coleman Ltd.

Also available in the David & Charles Series

THE SOMERSET & DORSET RAILWAY – *Robin Atthill*
BUYING ANTIQUES – *A. W. Coysh and J. King*
INTRODUCTION TO INN SIGNS – *Eric R. Delderfield*
THE CANAL AGE – *Charles Hadfield*
THE SAILOR'S WORLD – *Captain T. A. Hampton*
OLD DEVON – *W. G. Hoskins*
LNER STEAM – *O. S. Nock*
OLD YORKSHIRE DALES – *Arthur Raistrick*
THE WEST HIGHLAND RAILWAY – *John Thomas*
BRITISH STEAM SINCE 1900 – *W. A. Tuplin*
RAILWAY ADVENTURE – *L. T. C. Rolt*

The David & Charles Series

The Countryman Wild Life Book

Edited by
BRUCE CAMPBELL

UNABRIDGED

PAN BOOKS LTD : LONDON

First published 1969 by David & Charles (Publishers) Ltd.
This edition published 1971 by Pan Books Ltd,
33 Tothill Street, London, S.W.1.

ISBN 0 330 02740 9

Printed in Great Britain by
Richard Clay (The Chaucer Press), Ltd, Bungay, Suffolk

Contents

Foreword   *page* 14

Part 1: In the Field

Moonrise, by Sacha Carnegie, illustrated by Donald
  Watson   18
Island Pond, by D. R. Saunders, illustrated by Robert
  Gillmor   21
Snakes Alive, by G. E. Wells   24
After-Dinner Experiment, by the Earl of Cranbrook   27
Mounted Observation Post, by Henry Tegner   29
Saved While Serving, by Mary Hart   31
Swallow Revived, by Winifred Snape   32
Keeping Warm, by Bernard G. Price   32
Crossbill Tragedy, by R. D. Humber   32
Undaunted, by L. P. Renateau   33
Redstartled, by M. C. Adams   34
The Wrong Tree, by A. L. Durand   34
Off with the Old, by F. Miller   35
Sins of Omission, by Maxwell Knight   35

Part 2: In the Hand

Tits and Bullfinches, by Len Howard, illustrated by Donald
  Watson   39
Winter Friendship, by John McCaffrey   42
Palmed, by R. P. Gait   43
Faithful Fieldfare, by H. O. Ventress   43
Love from a Stranger, by Keith Hamilton-Price   45
Three Buzzards, by H. G. Hurrell   45
Aggression at Hand, by J. S. Beswick   50
Two Tame Ones, by Lolita Alexander   50
Jackdaw Jill, by W. Hacker   51

Combined Operation, by J. P. Gordon *page* 52
Voles on Pipette, by Andrew Pearson 53
Almost Non-Stop, by Stella M. Turk 53
To Tame a Vixen, by Vivienne Letts 54

Part 3: Hunting and Feeding

The Spitting Spider, by Stephen Dalton 58
Ground Sheet, by J. A. L. Cooke 59
Fungus-Eating Fly, by C. B. Williams, FRS 60
Toads and Bees, by J. H. Tullis, Thomas Watson, R. W.
Brown, S. E. Alsop and W. H. Tarn 61
Bee Feast, by Doris Wright 62
In the Raw, by Stephanie J. Tyler 63
The Gourmets, by Margaret B. Kabell and Dorothy
Spearing 63
Wormed Out, by S. Winifred Harper 64
Fly-Catching Fox, by Basil Cullin 65
Off the Glass, by W. T. G. Boul 65
Food in Comfort, by Joyce Fussey 66
Using the Wind, by R. T. Shepherd 66
Your Turn, Jack, by Sheila Harries 66
Food in Winter, by Winwood Reade, Arnold Bosworth and
Nancy M. Kelcey 67
Adaptable Swallows, by Heather M. Rudkin, K. E. Pearce,
Patricia Veall, J. S. Taylor and Emily Manini 68
Cyrano, by Freda Hurt 70
Great Grey Shrike, by H. R. Tutt 70
Dance of Death, by Anne Montrose 71
Winner Lose All, by H. A. Molloy 72
Opportunists, by A. B. Phillips, H. Batterbee, R. M.
Thompson and Joyce Hardiman 72
Danger at a Distance, by Mark Taylor 73
Stoat at Table, by S. R. Davidson 75
High IQ?, by Richard Tinklin and J. W. Gittins 75
Adaptable Finches, by Janet Kear 76
Reasoning Crows, by F. C. W. Stevenson, Frank C.
Edwards and C. J. Jacobs 77
Intelligence Test for Sparrows, by F. Pitches 78

Part 4: Confrontations

Slow-Worm Duel, by Violet I. Ricketts   *page* 81
Blown up for Safety, by J. A. Tudge   81
Cormorant Retaliates, by John Peterson   82
Little Owl v. the Rest, by J. W. Fishwick   82
Pheasants in a Garden, by E. Faithfull   84
Stop Go, by Stella V. Mann and G. R. Oliver   86
Giddy-Go-Round, by Stewart Campbell   87
Not Impressed, by J. T. McLean   87
Hickory Dickory Dock, by Elizabeth W. Douglas   87
The Best Defence?, by James Ker Cowan and Alan S. Bevan   88
Buzzards in Trouble, by D. McHattie and E. Anderson   89
Squirrel and Stoat, by G. A. Lovenbury   90
Terrier Rabbit, by D. Knowlton   90
Not So Funny, by Florence Hopper; comment by Winwood Reade   91
No Hiding Place, by J. Dearnley   91
Tug of War, by Brian W. Burnett   92
Stoats at Work, by Raymond Hewson, illustrated by Donald Watson   92
Tooth versus Claw, by Desmond Hawkins, J. E. Hemingway, P. Hale, R. J. Jennings and S. C. Marshall   95

Part 5: Routines and Displays

Birds at Play, by T. J. Richards, illustrated by C. J. F. Coombs   99
Image-Fighting, from contributions by H. J. Vosper, Molly Sole, Betty Cohen, Brian W. Burnett, Mollie Ingram, W. H. Archer, Mary Rickman, and Chris Mears; comments by D. W. Snow and Derek Goodwin   103
Ants but no Anting, by Hans Löhrl   106
Follow My Leader, by Greta D. Phillips   106
Birds Sunbathing, by C. H. Cooke   107
Pigeon Patrol, by P. G. Brade; comment by Derek Goodwin   108

Grandfather's Way, by George H. Hamilton; comment
  by Niko Tinbergen, FRS   *page* 109
Murmuration of Starlings, by Elizabeth Middleton   110
Labour-Saving, by John E. Platt   111
Minuet, by Janet Bodman   111
Rough Courtship, by Henry Tegner, illustrated by Donald
  Watson   112

Part 6: The Crisis of the Year

A Mother's Shield, by Ray Palmer   115
Stolen Silver, by A. C. Hilton   115
Trial Run, by Kay Shimmer; comment by Monica Shorten   116
Midwife to a Mole, by Frank G. Reeman   117
Badger Nests Above Ground, by Ernest Neal   118
Impatient Dipper, by Frank Bodman   121
Short Back and Sides, by R. Harrison   122
Fresh Feathers, by H. Stanley Jones and M. Ripley   122
Temperature Control? by Helen Robbins   123
Ducklings' Descent, by Charles Sherdley   123
Family Fellowship, by Anne L. Cooper   124
Seven to Safety, by Fred J. Chapple   124
Weasel Rescue, by Ann Blakiston   125
Child Care, by Marjorie Holland   126
Sitting it Out, by Bobby Tulloch   127
Through the Storm, by Phillip Glasier   127
Shrike Curiosity, by R. P. Gait   128
Multiple Nests, by Bruce Campbell   128
Two Cuckoos, by P. J. Lennon   129
Fostered by Flycatchers, by Doris E. Theak   129
Laid Out, by Florence E. Pettit   130

Part 7: Scale-Wings

Swallowtails for Wicken, by Brian Gardiner   132
Butterflies from Spain, by L. Hugh Newman   136
Butterfly in the House, by Ida Lumley   139
Purple Highflyer, by A. M. G. Campbell   140
Moth Camouflage, by M. W. F. Tweedie   140

Part 8: Mainly Mammals

Fallow Deer, by P. H. Carne, illustrated by Eileen A.
  Soper   *page* 143
Busby and Friend, by Gwendolen Barnard   147
Field Mouse's Tea-Party, by Alan Langman   148
Shedding the Load, by C. K. Mylne   148
Brock Gated, by Hilda M. Taylor   150
Coypus in Norfolk, by J. J. Buxton   150

Part 9: Various Voices

Wart-Biter, by John Burton   154
Starling Virtuoso, by P. M. Driver   155
Frogs' Screams, with contributions by John Usborne and
  B. K. Allen   156
Voles Taped, by John Burton   158

Part 10: Flora's Realm

Northern Orchids, by Alex MacGregor, illustrated by the
  author   160
The Whitty Pear, by Augusta Paton, illustrated by the
  author   164
Winter Fare, by P. R. Ivens   167
The Way of It, by L. R. Smith   167
Close Harmony, by Lilian Devereux   168
All from Bird Seed, by David McClintock   168
Tree with a Future, by G. H. Knight   171

Part 11: In the Water

Fish-Watching, by Colin J. Martin   176
A Movement of Bullheads, by E. A. C. Husbands   181
Slugs Underwater, by Joan Blewitt Cox   181
Grip on Life, by Donald McGregor   182
Ladies Only! by Bobby Tulloch   182

Coming in to Roost, by Susan Cowdy, illustrated by Robert Gillmor *page* 183

Birds Bathing, by E. M. Barraud, illustrated by R. A. Richardson 184

Magpie's Swim, by Hazel Inglis 188

Dew for Two, by A. T. Hitch 188

Owl's Bath, by M. Littledale; comment by H. N. Southern 189

Aquatic Deer, by Henry Tegner 190

Assisted Passage, by Cecil Nurcombe 191

Mink at Large, by Penelope Cook 191

Otters at Play, by T. R. Barnes 192

Otter and Dogfish, by Andrew G. Neal 193

The Blind Lady of Little Skerry, by R. M. Lockley 193

Index to Plants and Animals 197

Index of Contributors 203

List of Illustrations

(*between pages 80 and 81*)

A great tit lands
(*By courtesy of J. S. Beswick*)

Aggression wins a great tit a nut
(*By courtesy of J. S. Beswick*)

Heron in a Wexford garden
(*By courtesy of Maeve O'Keefe*)

Two mistle thrushes
(*By courtesy of David Alexander*)

Vole on pipette
(*By courtesy of Andrew Pearson*)

Vole feeding on a nut held in the hand
(*By courtesy of Andrew Pearson*)

Stone curlew with newly hatched chick
(*By courtesy of Phillip Glasier*)

Stone curlew after half an hour in a thunderstorm
(*By courtesy of Phillip Glasier*)

Birch-leaf shieldbug
(*By courtesy of Ray Palmer*)

White silk on filter bed
(*By courtesy of Andrew T. Paton*)

Hen house sparrow feeding on a tray
(*By courtesy of F. Pitcher*)

Hen pied wagtail posturing
(*By courtesy of H. J. Vosper*)

Hen pied wagtail flinging herself against mirror
(*By courtesy of H. J. Vosper*)

Hen pied wagtail gazing from top of mirror
(*By courtesy of H. J. Vosper*)

Jay in full anting posture
(*By courtesy of Hans Löhrl*)

Starlings sunbathing
(*By courtesy of the late C. W. Teager*)

Newly born mole
(*By courtesy of Frank G. Reeman*)

Badger cubs on Somerset farm
(*By courtesy of Ernest Neal*)

Lapwing on nest in snow
(*By courtesy of Bobby Tulloch*)

Red-backed shrike at nest
(*By courtesy of R. P. Gait*)

Spitting spider
(*By courtesy of Stephen Dalton*)

(*between pages 128 and 129*)

Our largest and rarest bush-cricket
(*By courtesy of E. Schumacher*)

Swallowtail caterpillars on carrot
(*By courtesy of Simon Frey*)

Swallowtail butterflies pairing at Wicken Fen
(*By courtesy of Simon Frey*)

Adult Swallowtail laying eggs on fennel
(*By courtesy of Simon Frey*)

Peppered moth and its black variety on sooty bark
(*By courtesy of M. W. F. Tweedie*)

Peppered moths on lichen-covered bark
(*By courtesy of M. W. F. Tweedie*)

The puzzled thrush
(*By courtesy of Ernest Neal*)

Hedgehog and parasites
(*By courtesy of C. K. Mylne*)

Coypu with its head in water
(*By courtesy of J. J. Buxton*)

Coypu in profile
(*By courtesy of J. J. Buxton*)

Red-necked phalaropes
(*By courtesy of William S. Paton*)

Owl's bath – on the brink
(*By courtesy of M. Littledale*)

Owl enjoying the dip
(*By courtesy of M. Littledale*)

Owl before the dust bath
(*By courtesy of M. Littledale*)

After the dust bath
(*By courtesy of M. Littledale*)

Otter and dogfish alerted
(*By courtesy of Andrew G. Neal*)

Otter wrestling with tough-skinned fish
(*By courtesy of Andrew G. Neal*)

Otter pops up
(*By courtesy of Andrew G. Neal*)

The Blind Lady and her calf
(*By courtesy of R. M. Lockley*)

Foreword

This selection of natural history articles, notes, drawings and photographs from *The Countryman* is made up almost entirely of items from the forty numbers published in the ten years 1959–68 and, though I set out with no such intention, every one of the forty issues is represented by at least one article, note or illustration; the book may therefore reflect what has caught the interest of the naturalist and country-lover during a decade when awareness of the importance of our wild life and of the threats to it was growing rapidly.

The task of selection was formidable because any note which is printed in the magazine's features 'Incidents of Bird Life' and 'Wild Life and Tame' is the survivor of about fifty offered. To make my task a little easier I have excluded notes about domestic animals, except where they were involved with wild ones, and almost all contributions from overseas, believing that readers of this book will be interested primarily in the familiar wild animals of Britain and Ireland which are probably better loved and more thoroughly studied than any other fauna in the world.

So do not look in these pages for more than the occasional rarity: some fifty kinds of bird and thirty of mammal are the chief actors, because they continue every year to produce traits of behaviour to delight the amateur and surprise the expert. Experts cannot be everywhere and frequently the man who sees something unusual as animals hunt, feed, fight, display, build nests and care for their young, is on the spot because of his work in field or forest; but he may be a business man with a moment to spare and binoculars in his dashboard pocket or a holidaymaker camping by a Scottish loch. Many of the keenest and most patient observers are housewives, looking out of their windows at the visitors to the bird table, or working in the garden while the dinner is cooking. The chance of seeing something that no one has recorded before is the excitement of natural history today, replacing the old collection of specimens; and the pages of *The Countryman* have always been full of such first-hand accounts.

Sometimes this has led to an advance in knowledge, such as resulted from the correspondence about frogs' screams summarized on page 156. These letters were invited, but the group of notes on the competition between predatory birds and mammals (see page 95) were offered quite independently. I think it can be argued that if such fleeting incidents are seen by a few people in different parts of the country, then they must be occurring unobserved quite regularly. Image-fighting on the other hand (see page 103) has a much better chance of being recorded, and it is time someone used the observer network of the British Trust for Ornithology to find out more about its incidence; at present our information is mainly anecdotal. Is it, as present knowledge suggests, confined like anting to passerine birds?

The articles and notes are grouped into eleven parts of varying length. Many items could find a place under more than one heading, but these divisions give a rough plan to the book, which is essentially one for dipping into. Each part has a short introduction and some of the notes have attached comments either by me or by one of the experts who advised us on different subjects. But in general our contributors speak for themselves, as well they may, for they include many who have been in the forefront of field natural history over the past twenty years and have done much by their writings, broadcasts, drawings and photographs to popularize the study of wild life and so help form a public opinion favourable to its conservation.

Others may not have been so prominent but show their interest and knowledge by sending items to *The Countryman* from time to time; and there may be some whose first printed words are reproduced here. Including artists and photographers, some two hundred contributors, with home addresses from Shetland to Cornwall, have helped to make up this book.

Burford, November 1968   Bruce Campbell

Part One

In the Field

Introduction

Most readers of *The Countryman* look at animals for the simple pleasure that it gives them, though they may have aesthetic and scientific reasons as well. The aim of this first Part of the Wild Life Book is to set the scene for the rest by giving examples of these approaches as expressed in articles and notes; some notes describing incidental relationships between watcher and watched are included here: those of a more lasting kind are the subject of Part Two.

The first two articles combine the aesthetic and scientific approaches with a large measure of enjoyment. Sacha Carnegie, who farms in Aberdeenshire, is an accomplished writer, serious and humorous, with the power to evoke the Scottish scene he knows so well. David Saunders went from the Cotswolds to be the first warden of the Skomer Nature Reserve; he is now organizing Operation Seafarer, which aims to define more accurately than ever before the location and size of our sea-bird colonies. In lighter vein G. E. Wells is embarrassed by her cats and their assiduous snake-hunting; and Lord Cranbrook explodes a hard-dying fallacy. Henry Tegner, a leading authority on the roe deer and on Northumbrian wild life, sets out the advantages of horseback as a mobile watch tower.

Four short notes illustrate the resourcefulness of bird-watchers in various crises; and A. L. Durand must have a special attraction for animals: another note of his told how a starling on a desolate Norfolk beach took refuge in his clothes. Finally, in the last piece he wrote for *The Country-*

man, Maxwell Knight unfolds with characteristic modesty some lost opportunities, and points a moral for all who go into the field with their eyes open and their notebooks shut.

MOONRISE by Sacha Carnegie

In the early dusk I went to watch for roe. There was snow at the back of the wind that soughed softly in the larches and rattled the tinfoil leaves of the rhododendrons. The ground was iron, the grass crackled, everything had the thin brittle feel of bitter cold. Soon the moon would rise; possibly the woodcock would come in.

I had watched the comings and goings of the roe deer on many occasions and knew the best places to wait. I chose a little clearing below a slight rise, along which grew a narrow belt of larches. Beyond the trees the glow of the hidden moon tinged the sky with red. The last remnants of daylight died and all colour faded from the trees, turning them to flat black silhouettes. A deer path wound its secret way through the larches, passing a few yards from where I stood blowing on my fingers. If the roe were going to use that path I hoped it would be soon. The top of the moon

edged into the frozen sky, rising slowly out of the sea. The evening was very quiet; the wind had dropped to a rustling whisper; a blackbird *chink-chinked* and a distant owl uttered its first mournful cry.

Then quite suddenly out of the dusk the redwings came. The air was astir with small black shapes flying along the widening track of the moon, pouring into the trees all round me. In their hundreds and their thousands they came, on and on, without a pause: an unending stream, silent save for the faint whirring of countless wings. They dropped into the branches and set up a great chittering and chattering; a clamour of whistling and piping and clapping of small wings. The whole wood became alive with the sounds of their settling down. Then, as though controlled by some central intelligence, the birds fell silent, all together, just like that. For a while they were restless, springing from the branches with terrific clatter and commotion, milling madly in the moonlight before settling back again.

An owl came sweeping on the icy wind, flying low above the trees on silent wings; a few little cries of sleepy alarm were raised, but not a redwing moved. Disappointed at their lack of cooperation, the owl vanished to search for mice and beetles in the silver grass. Somewhere deep in the wood a rabbit screamed, but the terrified cries were cut off abruptly. I shivered and longed to stamp my feet, swing my arms, but if you wait for roe you hardly dare to breathe.

The first appearance of the deer could not have been better arranged, not for the finest of nature films. At the very moment when the moon rose clear of the distant sand dunes they stepped daintily out of the shadows and into the brilliant light: two small deer, a buck following a doe. As though aware of their beauty, they stood motionless in the middle of the shining moon, not more than ten yards away. If they had been cut from hard black paper they could not have been more perfectly etched: a living woodcut framed by a fine tracery of gently waving branches and dark tree trunks. I could see the little antlers of the buck and his twitching ears.

Slowly I inched closer, silently on a bed of pine needles, slightly below the animals and looking up. The buck was feeding contentedly, but his companion was uneasy, jerking her head up, advancing a short step, stamping with her

forefoot, peering into the dangerous shadows, sensing something and not sure what. With a long stick I could have scratched her back. Beyond her clouds were stretching for the moon, outriders of the coming storm. Thin hard snow rattled in the branches, and the sound of the wind took on a higher note. But still the buck went on feeding.

The light began to dim. His outline was not so clear, the details blurred. I took out my handkerchief and waved it quickly. Two heads came up as one. For an instant the little deer was frozen. Again I flicked the handkerchief. This time the buck came forward to investigate the strange white movement, moving stiffly on springy legs. For an instant the clouds cleared, and the light behind him was so brilliant that I could see the grass sticking out from his mouth and the breath steaming at his nostrils. I flattened close to the ground, keeping my face lowered. I am sure that, if the squall had not swept between us, he would have come within reach of my hand; but the grey curtains of snow drove swiftly across the moon. All at once the deer vanished in the sudden swirling darkness.

The wind rose in fury, sweeping the blizzard through the swaying, dancing trees, and I knew there would be no more watching that night.

ISLAND POND by D. R. Saunders

Most of us have a favourite pond, and mine is a little stretch of water on Skomer, off the coast of Pembrokeshire. It has a charm all of its own. No trees fringe it, no farm ducks paddle on its waters, no fish swim in it; for a dragon-fly to skim over it is a rare event. It covers about half an acre and, when full, is not more than three feet deep; in summer it may shrink to a puddle and has been known to dry out completely. At one end a few stunted bushes of willow and sallow cover the small embankment which encloses its waters; through a gap in the dam a tiny stream gurgles its way to the sea half a mile off. The wild iris enriches the June scene with its yellow flowers along much of the banks; and on one side a patch of heather makes a splash of colour on dull autumn days. Great tussocks of sedge and rush hide shallow pools where water mint, marsh pennywort, cuckoo flower and bog pimpernel grow.

So small and open, the pond has many moods, quickly reflecting the vagaries of wind, sun and rain. On clear spring days it gleams an azure blue, but this changes to a muddy green in times of summer drought. When gales pile huge seas against the western cliffs of the island, the surface is whipped into wavelets which lap the embankment, and the wind whines through the bushes, breaking the iris down in its fury

Many birds are attracted to feed, rest, bathe or preen at the pond. In spring the sun gleams on the bottle-green heads of the mallard drakes, while their mates sit on nests in the bog below. Later the ducks shepherd their broods along the water's edge, keeping near cover in case of attack

from above. In autumn and winter small parties of teal
dabble in the shallows, joined occasionally by wigeon. I once
saw a shoveler drake and two female goldeneye. Usually
a pair of moorhens nest there, but they are rarely to be
seen, and often the first sign of their success is the appearance
of the young at the pond, their behaviour being quite unlike
that of mainland birds. The heron is an occasional visitor
but does not stay long, as the gulls mob it unmercifully.

My favourites are the true waders, which pay their fleeting
calls when the great autumn movement is in progress. Wood,
green and common sandpipers have been there – once all
three together. Occasionally a greenshank flies off as I
approach, and returns after I have entered the hide we have
built beside the pond. Little stints have several times rested
and fed by its waters; hardly bigger than house sparrows,
they show no fear of man. Ruff, redshank, dunlin, lapwing,
snipe and jack snipe are other visitors and once, after a
period of gales, a grey phalarope spent the day swimming
and spinning on the surface.

The pond is much used by gulls: some great black-backs but mainly herring gulls and lesser black-backs, which have a large colony less than two hundred yards away. All day in summer a stream of gulls comes to bathe and preen; sometimes sixty may be there together. The air is full of the noise of their splashing and flapping, and the green

water runs in droplets from their white heads and necks as they bathe. When the brown juveniles arrive in July they fight among themselves, and many of the weaker ones are killed. The water recedes, leaving the bodies, which are greedily eaten by their relatives and by foraging ravens and crows. One of the victims, an immature great black-back, had been ringed on Fair Isle. Though nesting in thousands on the cliffs, the kittiwakes seldom come to the pond; but one spring a pair of black-headed gulls, which do not

breed on the island, raised our hopes temporarily by taking up a territory nearby and driving off all avian intruders.

The willow bushes on the embankment shelter many small migrants. Occasionally, in spring, willow warblers and chiffchaffs sing a few snatches which seem to bring a breath of woods and hedges to Skomer. In autumn a robin usually claims a territory and shows its flame-coloured breast among the withering leaves. The wheezy call of a reed bunting and the scolding *tak-tak* of stonechats are sounds I associate with the pond at all seasons. Often rock and meadow pipits, with pied wagtails which nest in the ruined farm and, rarely, a grey wagtail, feed on the flies attracted by the warm mud when the water is low. Best of all the small birds are the swallows and martins. When, on a still evening in early April, I see the first of them swoop low after insects or to sip the pond's waters, I feel that spring has really come to the bare island.

SNAKES ALIVE by G. E. Wells

You may think there is little chance of your being able to study our native snakes at close quarters. I did until I moved to a remote spot in Hampshire with my two Siamese cats and a half-tabby. Their delight thereafter was to catch and bring indoors snakes of all descriptions and sizes, and for the most part very much alive. The cats seemed to think I should be delighted too, but I did not enjoy coming in hot from the garden with the prospect of flopping into a chair on top of an adder, or of thrusting my foot into a shoe which had become the temporary refuge of a baby grass snake. They might get anywhere, even into the bed, after the cats had become bored and left them to their own devices.

It was at breakfast that I first knew my scattering of snake knowledge was to be enlarged. I was peacefully eating bacon and eggs with my mind on pleasant things, and not even the arrival of Grandma Siamese scuffling and puffing attracted my immediate attention. When I did glance down my heart missed a beat. Her brilliant periwinkle-blue eyes gleamed triumphantly up at me over the wickedly glittering eyes of a snake which she held firmly by the back of the neck. To have a better tell-all-about-it she dexterously let go and sprang back. The snake instantly coiled itself into what I imagined to be a striking position with its head reared and weaving from the centre. All the oddments I knew about snakes began to flick in and out of my mind as fast as the creature's long, forked, black tongue flicked in and out of its mouth. I knew that there was only one poisonous species in Britain, the adder, and thought I would recognize it; but, confronted with this one, I hadn't a clue. It was certainly 'a pretty snake with a V on its head'; and it looked vicious enough. My bare sandalled feet seemed very vulnerable.

The only method my startled memory could suggest for dealing with a dangerous snake was to keep still, to play music on a pipe and to wait until someone chanced along to blow off its head with a gun. I was expecting the baker in an hour or so, but I doubted whether he would have a gun; and we obviously could not go on glaring at each other all the morning. So, forcing myself to be practical, I reached slowly across for the fire shovel without moving my feet and, with a quick movement, pinned the snake down with the edge, breaking its neck. When its head went on biting at the shovel I put it in half a bucket of water. I then hunted up my only book with any reference to snakes. It was most inadequate. The adder, it told me, varies a great deal in colour. It is inoffensive if left alone, but foolish people often kill both it and the charming harmless grass snake in senseless panic.

I looked the cat over for swellings, hysteria and other symptoms attributed by the veterinary dictionary to snake-bite. She purred contentedly at the attention. Then I took her and the other two cats to the dead snake and slapped them soundly, saying 'No!' They departed much offended but with dignity unimpaired.

Later in the day, on coming in from the garden, I was quite unnerved by the sight of the hearthrug trying to get under the chair, apparently of its own volition. The cats, intensely interested and pleased with themselves, were poking at it experimentally. I did not need to glimpse the scaly bulge that writhed momentarily from under the rug. From its size I took it to be a large snake and therefore not an adder which, according to the book, rarely exceeds twenty-four inches. Fighting down my feelings of senseless panic, I resolutely seized the rug where I estimated the snake's neck to be and hurried a good two hundred yards down the lane. Even through the thick material the sinuous feel of it turned me over, and I was thankful to be able to cast the rug from me. The snake which slithered off into the undergrowth was not as long as I had expected and appeared to have a zigzag stripe. I wished I had taken it a bit farther and hoped that snakes did not have homing instincts.

The next day Cleo was making lightning jabs at a creature which I hoped was a grass snake, and the tabby had just the tail of another. Had she eaten the rest, or was it lurking somewhere at the ready? It was lizards, I dug out of my memory, which were unaffected by the loss of a tail. Nevertheless I carried the tail on a shovel to my neighbours, who all assured me that they would know an adder if they saw one; but they could not tell from the tail. It might well be an adder, they said; it had the orange-mauve underparts. If they could see the whole snake . . .

The cats caught various other specimens, but as long as they kept these outside I left them alone, except when I saw them peering intently into the undergrowth and felt an irresistible urge to join them. Sometimes it was a bee or a mouse. If it was a snake (and I could see it) I took the cats off and told them to leave it alone. One freshly cast skin or slough gave me quite a thrill, having no unidentified danger inside it. I acquired another book, which was slightly more enlightening than the first, and was surprised to find that there were only three kinds of native snake, apart from the blind- or slow-worm; the cats seemed to have had so many sorts, due (I gathered) to the fact that the colouring of all three kinds varied. The female adder illustrated looked to me like one or two I had come to regard as harmless grass snakes.

It was now my habit, when returning to the house, to take a suspicious look round; and a day or two later the smug expression of Grandma Siamese, sitting on the back of my chair, caused me to push the cushion forward with the poker. There sure enough was a nice specimen of a snake, about eighteen inches long. It lacked what one book described as the 'well-known markings of the adder'; nor did it closely fit any other description. Determined to have it properly identified, I pushed the cushion back over it and crossly shut the cat in the kitchen. I then cut a cleft stick – a miniature clothes-prop – and, armed with this and the long-handled shovel, returned to the snake. I got it on to the shovel, held it there with the fork of the stick behind its head and set off briskly in search of an expert. I was carrying it well in front of me as I went out into the lane, so the snake went first through the door and, as it happened, almost into the face of a youth wandering past. He turned the colour of putty – the new greenish putty – and I had to put the snake down carefully and at a little distance to go to his aid. When he had recovered the snake had made its escape.

By the time I came across another doubtful specimen I had a glass-sided case to confine it, and I carted it round. All the folk who had been certain that they would know one kind of snake from another, and an adder for sure, were now quite unsure. The books said that a grass snake could be easily tamed and made a charming pet, whereas an adder could not be kept in captivity, as it would starve itself to death. This one was not tempted by worms and a tiny frog during the night but, on a visit the following day, the vet pronounced it to be a grass snake.

'So you would not mind wrapping it round your wrist?'

'Not on your life,' he said, and the creature continued to starve itself until it was released. The vet also said that the latest census of adders in the area showed twenty-five to the acre, and that he had been called to forty cases of snake-bite in cattle, horses and dogs but never in a cat.

AFTER-DINNER EXPERIMENT
by the Earl of Cranbrook

When Mary came to dinner we had not seen her for some years. Shortly after the war she and a friend, both just out of the ATS, had come to work on one of the farms, partly with the cows and partly on a market-gardening enterprise which I had started at about that time. A pretty girl, she had grown into a beautiful woman with just a tinge of grey in the dark hair over a young face. During the early summer I had experimented with two eighteen-year-old blondes, and here was an older brunette to act as the control necessary in any carefully designed scientific investigation – and an animal-loving brunette at that.

It was a warm evening, like all the summer evenings of 1959, and the windows and the door were wide open, with the smell of tobacco plants drifting in. By great good fortune there came in, too, a pipistrelle which was shepherded into the outer hall and caught with a butterfly net. Struggling, squeaking and biting furiously, it was obviously a better subject for experiment than the tame one I had in a cage, so I lifted up a handful of Mary's curly hair and popped the bat in. I had done the same with four different species of bat and the blondes, and the earlier results were repeated: the pipistrelle scrambled up over the top of Mary's head without getting entangled in any way and took flight out of the door, back into the night.

It is, of course, almost axiomatic that women's hair has an irresistible attraction for bats and that, once in contact, the two become so inextricably entangled that they can be separated only with scissors wielded by a man. The stories of such occurrences are much like those of the Indian rope trick: one's informant knows somebody, who knows somebody else, whose first cousin knew a girl ... It is extraordinary how they grow. I was told by a descendant of a long line of country parsons – country parsons of the time when they and country squires had the education, leisure and inclination to indulge in scientific and literary pursuits – that there had always been a tradition in her family that some time in the middle of last century a bat flew into the hair of her great-aunt Nellie while she was sitting in front of the fire with her feet in a mustard bath. Mustard baths and bats in the house are, of course, almost mutually exclusive: the first is a by-product of November fogs, the second of high summer. Yet this story grew up in a family with a strong tradition of critical scientific observation. When bats are said to fly into the hair, reason flies out of the window.

I must confess to considerable vicarious pride that in 1959, the centenary year of the publication of Darwin's *Origin of Species*, three women from my own county were ready to offer themselves thus as martyrs in the sacred cause of science to test the truth of a hoary superstition. It is much to be hoped that some stout-hearted West Country women will be found to experiment with the greater and lesser horse-shoe bats, which are not found in Suffolk; we have only Vespertilionids.

MOUNTED OBSERVATION POST
by Henry Tegner

Only recently did I come to realize what a wonderful observation post a pony can be. I had been trying for some months to observe the activities of a small party of roe deer which had taken up residence in a twenty-acre wood at the back of the house. These deer can be extremely difficult to watch. They seem to prefer to move about in the early hours of the morning or late in the evening and, unless they are entirely undisturbed, do not show themselves much during the day. They are also essentially woodland animals, knowing only too well how to make use of cover. In winter, when there is little leaf on the trees and the frost has cut down the undergrowth, it is possible to see them more often than in the summer; but long-term study of their habits requires considerable patience and woodcraft.

I had nearly given up the idea of studying my particular group when one morning I went out riding on a little fell pony of mine. I made for the wood where the roe live, as it is a pleasant place at all times of year. During the autumn and winter the ground is boggy, but a native pure-bred fell pony can teach most humans anything they do not happen to know about bogs and bad going. I had scarcely got into the precincts of the wood when I noticed not fifty yards away the stilted rump movement of a deer. Pulling up my pony, who immediately dropped his head to nibble at the sparse grass under the trees, I sat as still as I could and counted five roe together in a little cluster. One looking like an old doe stood staring at us with ears flared and wrinkled her wet black nose, searching the air for our scent. Apparently she was satisfied, for she soon began to browse on honeysuckle shoots, and I was able to watch all five until they fed out of my sight.

Strangely, as one rides through the fields, domesticated animals often appear to take more notice of a man on horseback than do the wild ones. Sheep frequently bunch and run at the approach of someone mounted. Cattle are usually inquisitive and, after approaching the rider, will skip and prance in seeming play. Wild birds too, I have found, often take little notice of a rider. One day my pony nearly stepped on a sparrowhawk which was on its prey beneath a haw-loaded thorn bush. Fieldfares and redwings will stay feeding on haws, apparently oblivious of a rider by their hedgerow. In spring the tits, treecreepers and warblers seem not to mind the presence of a pony at all.

In our part of Northumberland there were a great many field voles last year; at times they were reported to have reached plague proportions. They are not easy to observe, and my best sight of one was, again, from my pony's back. I was walking him across a grass field when I noticed a small creature scurry into a tuft of common rush; so I stopped and waited. Soon one vole appeared and then another, apparently looking for something to eat. As I watched them I heard in the sky overhead the indignant cawing of a pair of corbies. They were climbing in wide spirals towards a hovering dot, too big for a kestrel. It turned out to be a buzzard, a rare visitor in these parts, which may well have been attracted by the vole invasion.

The secret of successful observation from a pony's back

is that the mount should be docile. Jet, my black fell gelding, stands to order if there are no other horses with him. So far he has proved himself an ideal observation post. Before breakfast on the morning on which I write I saw one of my roe again – a young buck in full winter pelage, his horns still in position. He let me come to within ten feet of where he stood nibbling at a silver birch. I was the first to move, and when I did so he walked quietly away without a sign of this timid creature's customary bounding gait.

SAVED WHILE SERVING by Mary Hart

After wild courtship chases which took the hen bird right into the house, a pair of grey wagtails nested in a hole in the wall close to our kitchen; it overlooks a fast-flowing Devon stream four feet below. I placed cotton-wool and bits of hair on the sill, and they were soon picked up and used. The parents took turns to sit on the six eggs, which all hatched. Then I noticed that one chick was being ignored by the parents and missing its share of food. It seemed weaker than the others and was soon pushed to the edge of the nest with its head hanging over the stream. I was busy with the dinner and, just as I was about to serve the sweet, glanced out of the window. Seeing the baby bird fall on some gravel in the stream-bed, I was out in a flash to rescue it and continued to serve the meal with its apparently lifeless body clasped in my left hand. As I was clearing the table, it suddenly lifted its head and gave a solitary cheep, though its eyes were still tightly closed. This happened again, so I put it in a fur glove on the rack over the cooker, where it revived. We fed it on a few flies and ants – a tricky job, as it kept on passing out. But at last its eyes remained open, and it sat on my hand. I decided to put it back in the nest, but my action alarmed the parents and precipitated the other chicks into the stream. We rescued them all and finally quietened them by keeping a hand over the hole. The parents resumed feeding and the brood fledged successfully a few days later, leaving the nest one after the other. But one, which I felt sure was the resuscitated bird, kept returning to a stone in the stream and calling; when I cheeped back, it replied and made no attempt to fly. It reappeared next day, after which we saw the brood only occasionally. The parents probably nested again higher up the stream.

Swallow revived

Watching the swallows over the River Dee at Corwen as they flitted in the sunlight or skimmed the water after some morsel on the surface, we noticed one bird which did not rise after its dive. The current bore it towards a bank of pebbles where we were able to retrieve it and lay it on the grass apparently lifeless, water oozing from its eyes and beak. I had a smelling bottle in my pocket and quickly applied the stopper to the bird's nostrils. The immediate result was a tremendous shudder and blinking of the eyes. A second application brought the swallow to its feet, and in a few seconds it was up and away to resume its hunting. – *Winifred Snape*

Keeping warm

Camping near the River Itchen, I looked out from my tent early one autumn morning to see a covey of partridges grouped round the embers of the cooking-fire. There were no food scraps to attract them; but the air was damp and chilly, and they sat so still that I was sure they were there for warmth. – *Bernard G. Price*

Crossbill tragedy

For some years a small colony of crossbills have been resident among the few scattered larch plantations on the Lakeland fells at High Borrans, above Windermere. Their favourite feeding place was a stand of particularly fine larches, the tallest of which reached ninety feet. A timber merchant bought these trees, and when felling started last summer the flock of about forty crossbills continued to feed on the topmost cones of the remaining larches, while men worked below. When one of the last trees had come crashing to the ground the woodman in charge was surprised to hear twittering among the branches and to find a pair of crossbills trapped in the foliage. He picked up the brightly-coloured male, which flew from his hand, but the female was so badly damaged that she had to be destroyed. – *R. D. Humber.* [One autumn I watched the felling of a large elm at Oddington in Gloucestershire. Five minutes or so later I saw a woodpigeon fly out of the tangled branches. – *B. C.*]

Undaunted

In our New Forest garden we had a decaying horse chestnut tree, from which we had removed the branches and then the bark, before creosoting the trunk. Last year, on May 18th, two men cut this tree below ground level and, when it fell, my wife saw a green woodpecker in a hole near the top. She covered the hole with a coat and then stuffed a handkerchief into it. The men cut the trunk in three pieces and placed the top section in an apple tree with well-splayed branches on the lawn. The bird stayed inside throughout, and in the days that followed many chips appeared on the lawn beneath the hole, now only seven feet from the ground. On June 1st we found two woodpecker's eggs under the apple tree; of one no more than the shell remained, but the other was intact and we returned it to the nest. My wife kept this under observation in my absence during the last two weeks of June, and the woodpeckers would look out to welcome her. On July 4th I saw two young. Eight days later they had flown, leaving no egg behind them in the nest. – *L. P. Renateau*

Redstartled

One November, when I had been bird-watching for three years, I heard that a black redstart was frequenting the station at Calbourne, about seven miles from my home at Yarmouth, Isle of Wight. So I cycled over at the first opportunity, hoping to add this scarce passage migrant to my life-list. But a search of an hour or so failed to produce the bird, and I returned home. About ten-fifteen that night, as I was reading in bed, I was subconsciously distracted by a large moth fluttering at the window. It made a surprising amount of noise, so I got out of bed, threw open the curtains, and the creature flew in. I could hardly believe my eyes: it was a black redstart. It flapped round the room, unable to find the window; but when I opened the door, it flashed through and was expertly fielded by my father in his cap. On examination, it proved to be a female or immature bird, whereas the one at Calbourne had been an adult male. – *M. C. Adams*

THE WRONG TREE by A. L. Durand

As I wandered through a beech wood on the Chiltern escarpment early last May, enjoying the rich contrast of bright yellow-green against the warm brown of the forest floor, I became aware of a slight tickling sensation at the back of my neck. I put up a hand to investigate and something whizzed round my collar below my right ear, ran straight across my tie and, before reaching my breast pocket, flashed round to face me, with dark quivering whiskers and soulful black eyes, as it clung somewhat apprehensively to my left lapel. I was gazing at a native dormouse, now by no means common and rarely seen in daytime. As long as I did not move, it seemed prepared to stare it out; but when I tried to catch it, it shifted with an amazing turn of speed to the small of my back. For a further five minutes we kept up the game, the dormouse moving when I did – once it nearly got inside my shirt – until eventually it shot down my trouser-leg and twinkled two or three yards to a medium-sized beech, to disappear through a fork ten or twelve feet up. My friend, who had been nearly doubled up with laughter, said that, disturbed by our approach, the dormouse had emerged suddenly from the thick carpet of leaves and, with instinctive reflexes too quick for selection, taken me for the nearest tree.

Off with the old

A smooth newt in my vivarium stood with its head raised, opening its mouth wide. At each yawn, the grey-white skin rolled back from its nose. When a tight roll ringed it behind the head, it began to rub each side alternately against a stone. Slowly the roll moved down; the newt paused to withdraw each leg in turn, as though from boots, before continuing to scrape itself. After ten minutes, when the rolled skin was within a quarter of an inch of the tip of the tail, it suddenly twisted round and seized the tip in its mouth – one tug, and the roll hung from its lips, then was swallowed. Another newt shed its skin under water, also by rubbing against a stone. It came off in pieces, which hung in gossamer tatters around it before they floated away. – *F. Miller*

SINS OF OMISSION by Maxwell Knight

I often wonder how many amateur naturalists reflect on the interesting things they have seen or heard in their wanderings and, for some reason, failed to record at the time – still less write up for an appropriate journal. The moralists are always telling us that 'confession is good for the soul', and it may well be in this context. Recently I have indulged in a period of reflection on my own shortcomings, weeping silent tears over intriguing observations I have made over the years. Some might have been of value, had I not been lazy or forgetful. Amateur naturalists are inclined to be a little superior about their amateurism, referring to their qualified and professional colleagues as hidebound, sceptical and too serious-minded. Even if there is a grain of truth in these strictures, we all too frequently forget the value of a disciplined training, part of which concerns the prime importance of noting down all incidents observed.

I was made to welter in self-reproach the other day when a friend who knew I was interested in hedgehog behaviour asked me if I had ever seen one walking in circles, apparently aimlessly, and keeping it up for some time. I replied rather casually that I had, and thought no more about it until, a few days later, I was brought down to earth with a bump. I received a letter from a distinguished scientist who is a student of animal behaviour. He told me that my friend had passed on my remarks about the hedgehogs; and he asked for full details, with any references to this curious phenomenon I had read. I hung my head in shame when I realized that, though I had observed similar hedgehog perambulations on three occasions, I had taken insufficient notice and, worse still, had not written down the dates, times and other details at the earliest possible moment. I was able to give only such information as I could recollect. My memory is good; but that is not enough.

Another case also concerns hedgehogs. For some years Dr Maurice Burton has been investigating the curious self-anointing in which they indulge; and he was discussing it with me, describing his experiments with tame hedgehogs. He said he wished he could get information from someone who had seen this behaviour in the wild. I was at once reminded of the time, some years before, when I had without doubt witnessed it. But I then knew nothing of self-anointing and had taken the antics of the urchin for an ailment of some kind; as I had a dog with me, I was more concerned to keep him away from a possible source of infection than I was with the hedgehog's behaviour. Once again I was able to give only personal testimony without data of value.

Then there was the time I first saw a pile of dead frogs

near a spawning-pond, each with its hind legs clearly eaten by some predator. I did make a note of this, but failed to count the corpses and to see how many were males and how many females. Later, in discussions with Dr Malcolm Smith and Frances Pitt, I learned that this was the work of a stoat. Dr Smith wanted to know whether there were signs of eggs which had been left severely alone, and I had not troubled to look. I learned a lesson from this and have since seen a stoat, also a brown rat, at work of this kind; and I have heard of water voles behaving in the same way. Always the remnants of spawn have been quite untouched, probably because it is distasteful or even poisonous.

One of the most intriguing sights, which I also failed to write up fully, occurred when I was studying the marsh frog in the dykes on the Romney Marsh. It was after midnight, and the temperature was so cold that my companions and I were all wearing thick sweaters, mufflers and so on even though it was late May. We were using electric torches to spot the frogs in the water, which was equally cold. Suddenly, in the light-beam, three of us saw a grass snake in the middle of a dyke, its head raised, very alert and clearly on the hunt among the frogs. We caught the snake, but the interest lay in the fact that it was feeding at night and at a temperature which one would have expected to cause all good serpents to lie up in some snug spot on land. To be fair, we did make some notes; but, so far as I know, none of us wrote up the incident for publication in some suitable journal. On this occasion, I am happy to record, only one other amateur was present; all the rest were scientists of high degree, including a professor.

It is a little mollifying to reflect that even the mighty have their lapses. Even so it is good that all field naturalists should remember to record with as much detail and as soon as possible even those incidents which appear insignificant at the time. I wonder how many valuable records have been lost through our sins of omission?

Part Two

In the Hand

Introduction

A great many of the notes and articles offered to *The Countryman* are about pets, either the recognized domestic animals or waifs and strays from the wild. But the emphasis in this selection is on the wild rather than the tame, so W. Hacker's 'Jackdaw Jill' is the only one who comes in the pet category, and she nested each year with a wild mate in an unknown site. H. G. Hurrell's three buzzards are successfully returned to the wild in the Dartmoor setting we associate with this versatile naturalist-photographer and his family. Len Howard is well known for her detailed observations on mouth-tame garden birds, especially great tits, and her successful fight to save their sanctuary from the developer. She is perhaps the most gifted of a number of women who have been on close terms with small birds, and made valuable observations on their behaviour.

The selection of notes includes instances of different kinds of bird in relationship with man; some will find Maeve O'Keefe's heron the most remarkable, but apparently not without parallels, while I recall my astonishment when I saw one of Lolita Alexander's mistle thrushes come to hand. Keith Hamilton-Price's verger and his robin pose the problem of how wild birds identify people; we recently published another note about a robin which transferred from its original 'owner' to a relative of somewhat similar appearance; and there is some evidence that individual voices may be recognized.

In the items on mammals, the baby field voles reared by

Fenella Greig are 'tops' for round-nosed appeal; Stella M. Turk summarizes the high-pressure life of a shrew, and Josephine is an example of the most satisfactory relationship between a human and a wild animal, which abrogates none of its independence; but Vivienne Letts' warning about the dangers of hand-taming a fox is worth heeding.

TITS AND BULLFINCHES by Len Howard

An ever-increasing number of birds have inhabited my sanctuary since it was saved from destruction by builders through a public appeal a few years ago. In the winter months masses of hungry outsiders help the residents to consume many pounds of bird food each week. A score of coal tits take far more than their share for, like the marsh tits, they habitually store food, much of which they poke into the soil under bushes and plants.

Recently one of my marsh tits was on the bird table attached to the window-ledge, busily packing his beak with bits of peanut for his larder. The handsome bullfinch stood beside him, intently watching while the marsh tit made several unsuccessful attempts to hold yet another bit. At last he succeeded and, raising his head, looked up at the bullfinch, who gently took a piece from the overloaded beak, his polite manner suggesting, 'Let me relieve you'. The table was plentifully spread with nuts, so it was evident that the bullfinch was interested merely in the marsh tit's behaviour. Theft of food from a bird's beak by another species is quite usual if anything equally tempting is not easily obtainable; it is sometimes mistaken for intentional feeding of another species.

In summer this bullfinch often fed his young on peanuts and sunflower seeds from the bird table. One day when this was empty the young flew indoors to a dish of similar food where tits had been feeding. They examined this carefully and, not yet feeding themselves, flew to fetch father from a tree nearby. He did not follow them, although they repeatedly flew to him with hunger cries, then back indoors with call notes that grew persistently louder as their manner became more and more excited. Parent bullfinches, through man's persecution of them, have learnt caution; so I hung the dish just inside the french window. At once the father came to it, and the shouts subsided while he chewed the food, then fed the youngsters. Ever since they have called to me, as do many other species, if the bird table is empty when they are hungry.

Bullfinches have gentle ways towards smaller birds on the table and often step aside to wait for others to feed if it is crowded. My nuthatches behave very differently and seem

to enjoy clearing the table of every bird by darting right and left with thrusts of formidable beaks – an amusing sight when, in the nest season, aggressive sparrows are being swept off the table. I think this is done in fun; if really hungry, the nuthatch seizes food without disturbing the others and flies to a nearby oak tree with bark crevices suitable for wedging a nut, to be eaten by sharp blows. This method of eating needs practice by the young, and I have watched them make many unsuccessful attempts to wedge the food in unsuitable places; they then rush around with it, trying to find a fitting crevice.

Although I want to let all my trees grow undisturbed, occasionally those by the road need topping; and last September the Electricity Board sent a man with crane equipment to cut a tall macrocarpus that interfered with their wires. This troubled me, for the tree was of great importance to one of my great tits, Matey, and there had been close understanding between us since his fledgling days. His nest site was across the road, and this tree by my gate was his one piece of territory on my land, giving him rights to the cottage for food in the nest season. My sanctuary cannot provide nest sites for all the large number of great tits reared here and wanting to remain; so some of those without sites cleverly arrange to secure a tree or two on the boundaries of my land as, by a general understanding among the great tits, this entitles them to a right of way into Bird Cottage for food for their fledglings. Any who have not kept a footing in the sanctuary are regarded as outsiders and chased away in the nest season by those who have secured sites within it.

When Matey has young in his nest across the road, he continually flies to his macrocarpus, then indoors to get mealworms from my hand for them. In summer a continuous stream of the other great tit commuters fly along their rights of way into the cottage and back to their nestlings with beaks full of mealworms, some having nests at a considerable distance. At ruinous expense I have supplied mealworms since I discovered that great tits had serious difficulty in rearing young, most of their nestlings dying of starvation due to the extermination of their natural food by insecticides and tree spraying. So, before the cutting of Matey's tree began, I explained the whole situation to the man. He showed much interest, having read about my birdland. The cutting was done with thought for the importance of the tit's foothold in the sanctuary, the tree being carefully studied from over the road several times during the

process, to make sure that enough boughs showed from Matey's nest site for him not to be too upset.

In the mild days of mid-January my great tits were already beginning their intricate planning for the coming nest season. My two charming young bullfinches have had to nest elsewhere, but I know this will have been settled between them and their parents in a typically gentlemanly manner.

Winter Friendship

A young heron, believed to have come from a nest near the golf club, landed in the Misses O'Keefe's garden at Rosslare Strand, Co Wexford, in August 1965 (Illus. 3). At first it was weak and used to sit on the grass. Its food, fish or fat meat cut into strips, was put in a bucket of water. When fed and petted, it tucked its beak under its wing and fell asleep on its perch; it used to pirouette, apparently with pleasure. Gradually it grew stronger and was able to fly away to the slob-land, usually returning twice a day, until February 1966, when it disappeared until April. After a month 'in residence' the heron went off until November. In its second winter it was still quite tame and came to its bucket for food. It was not frightened of the two dogs; in fact, they were afraid of it. It would not come near if it heard strangers, but it recognized Maeve O'Keefe's voice and even accompanied her when she was playing golf. – *John McCaffrey*. [We also had a record of a heron which visited a Co Donegal garden for ten years. If the Rosslare bird was a female, she may have tried to nest as a one-year-old in 1966. – *B.C.*]

Palmed

My strangest experience with birds occurred after I had set out one evening along the River Wye with that fine photographer Arthur Brook in search of a sandpiper's nest. We were unsuccessful and took to the hills, where we saw a pair of sandpipers by a stream. As we watched from cover, the hen settled down on what turned out to be two chicks. It was then dusk, and the adults were so tame that my companion decided to experiment. Lying flat on the shingle, his cap pulled over his eyes, he held both chicks in the cupped palms of his extended hands. To my amazement the hen approached nearer and nearer until finally she brooded the young in Arthur Brook's hands, while the cock stood only inches away. – *R. P. Gait*

FAITHFUL FIELDFARE by H. O. Ventress

In the winter of 1954–5, when the fieldfares and redwings had finished the few berries in our Devon garden and moved on, one fieldfare remained. Between November and March it ate large quantities of rotting apples which, lacking other provender, we dotted about for it, hiding some under an old roof-tile from marauding starlings. The fieldfare became increasingly tame and would often perch in an apple tree only a foot or so from our kitchen window. When my husband said 'chock-chock' to it, the bird would reply. When it left in March we did not expect to see it again, but on November 13th a fieldfare alighted on our garden wall, then flew straight to the roof-tile and looked under it. My husband had put a few apples there on the chance, and the visitor stayed. In February 1956, the temperature never rose above freezing and about the middle of the month a pair of mistle thrushes moved in on us and began to attack the fieldfare. We would hear it squeaking for help several times daily, as the thrushes flew at it, rolled it over in the snow and pecked viciously at it, bringing out beakfuls of down which it could ill spare. We were constantly dashing to the rescue, till we caught the thrushes with a cinder-sifter propped on a stick and baited with an apple. Putting them in a box, in a biting east wind we took them some distance to a sheltered lane, and when we released them they flew away rapidly still farther from home. Unmolested, our fieldfare soon regained its strength, and when it left us in March it was quite restored. On November 23rd we again saw a fieldfare drop down on to our wall and go to the roof-tile for the apples that awaited its return. In the months that followed, it conducted vigorous sorties against three resident pairs of blackbirds, and its mellow 'chock-chock' resounded from earliest light till shortly before dusk, as it defended what it obviously regarded as its territory. Its vitality seemed undiminished in its third winter with us; and it arrived for its fourth season in 1957 and stayed until late February 1958; but we saw nothing of it during the next three comparatively mild winters and concluded sadly that it had reached the end of its span. Our joy may be imagined, therefore, when it reappeared in November 1961 in splendid condition. Soon it was responding as before, with a melodious 'chock-chock', to our attempted imitation of its call. It even remembered the red roof-tile under which, each autumn, I had placed a few apples in hope of its return; and though at first it found the hole empty, blackbirds having forestalled it, this was soon remedied. It could keep at bay our four resident pairs of blackbirds, but I again had to trap one particularly aggressive mistle thrush with a baited cinder-sifter and release it some miles away among well-berried bushes.

Love from a stranger

Mr Jenvey, the verger of St Nicholas, Brockenhurst, has in twenty years tamed many birds by a method which he calls 'cheese and kindness'. One particular pet is a year-old robin which comes to his call and will take cheese from his lips. In January we set up our BBC sound camera in the churchyard, screened by bushes, hoping to record the performance. The bird appeared every time at Mr Jenvey's call and perched on one of the gravestones only four feet away, but to his consternation it refused to fly up and take the cheese. It was not embarrassed either by our presence or by the camera, and appeared completely fearless but, though we waited two and a half hours, it would not go within four feet of the verger. Puzzling over the reason, I noticed that Mr Jenvey was dressed in a brand-new light grey-blue suit, a white shirt, a red tie and a new light-coloured trilby hat, and asked him jokingly whether he normally wore his Sunday best to feed the robin. 'Oh no,' he replied, 'I always wear my working clothes – old dark jacket and brown hat.' With his permission I drove to his home and brought back the old black reefer jacket and the battered dark trilby. He changed into them and called the robin, which came straight to him, took the cheese from his mouth and flew off before we could even reach the camera. Later it repeated the performance for our records. – *Keith Hamilton-Price*

THREE BUZZARDS by H. G. Hurrell

When by some misfortune a wild bird is reduced to a condition so helpless that it can be picked up by hand, its prospect of survival is frequently poor. How often has such a bird been given every care and attention, only to disappoint the anxious rescuer by its failure to recover! Such experiences, several times repeated, incline one towards a put-it-out-of-its-misery policy. Yet hope is not easily abandoned, and we are naturally reluctant to rule out too hastily the possibility of recovery. Our success, as a family, in resuscitating no less than three buzzards in the past two years has encouraged us to take quite an optimistic view of the recuperative powers of this species. All three were completely helpless when brought to us, though they were full-grown birds. They were picked up by hand – an indignity which no healthy adult would willingly tolerate.

The first arrival was a female, carried to the house by two boys who had found her on the ground almost unable to move. She must have been close to death by starvation, though the cause of her extreme emaciation was not clear. Fortunately, my daughter Elaine was at home and fed her without delay with pieces of raw meat pressed down the gullet, because she was too weak to feed herself. We anxiously looked in her box each morning, and it was three or four days before she was standing up. Her strength and bearing then improved rapidly, and some idea of her progress can be gained by her astonishing weight increase from seventeen ounces to twenty-nine ounces in the first six weeks. We then considered the advisability of releasing her. Doubtful whether she could become self-supporting immediately, we attempted to 'hack her backwards', as falconers would say. Food would be put out daily at the place where she was given her freedom. As long as she returned to feed she would be at hack; but if she replaced this diet with a natural one she might go back to a completely wild state and give up her visits.

I believe that young hawks, for some days after flying from the nest, possess an instinct to return to its immediate vicinity to receive the food the parents continue to bring. It is therefore fairly safe to hack a hawk as soon as it can fly. Later this instinct seems to diminish, so that attempts at hacking may fail. In view of this the prospects for our buzzard could hardly be considered favourable. Yet we wanted her to be free and were anxious to give her the best possible send-off. For several days she was tethered with jesses and leash to a bow-perch on the lawn and fed there. One or two large boulders near the house seemed ideal feeding stands for a hawk; so a line (creance) was attached to her jesses and she was encouraged to fly ten or twenty yards to these stones for a reward. Then one evening, after a full crop, the leash was removed and she was left free on her perch. That night she roosted in a nearby tree.

Next morning the weather was atrocious with mist, rain and a high wind. She was found in a tall ash a couple of fields away and enticed down to a piece of meat. The following day this was repeated elsewhere; but we did not seem to be training her to return to a particular hacking place. On the fourth day after release she was back, almost accidentally it seemed, to a tree within sight of the stones, which were at once garnished with red juicy meat. Anxiously watching from the house, we saw her fly a hundred yards to the food; and once this initial step had been taken she returned without fail daily. She was soon so strong on the wing that she was as competent in flight as the wild buzzards with which she sometimes soared. Yet during the whole of 1958 she scarcely missed a day, taking stand each morning in a favourite pine tree near the house till food was put out for her on one of the boulders. She would then glide swiftly down and, with precise timing, sweep the meat off the stone and rise with it in her talons. She usually carried it up to the branch of some tree, where she evidently felt safer than on the boulder.

Although our Devon buzzards are usually regarded as sedentary, I feared that, after nearly a year at hack, migratory restlessness might take ours out of range. On several occasions in September, on fine days, I saw her soar to immense heights and drift away over the moor as a speck in the distance; but next morning she would be occupying her usual perch. This was no small relief, because I could not forget the young buzzard which I ringed on Dartmoor some years ago as a nestling, and which was picked up dead in its first autumn near Newquay, fifty-five miles west-south-west of the place where it was ringed. Also, she might well disdain the meat provided, now that she was hunting on her own to a considerable extent. I once saw her pounce on a field mouse, and her interest in our pond when frogs were available had an obvious purpose. At one period she developed the habit of making sudden attacks on wood pigeons: seeing one sitting in a tree, she would fly straight at it, causing it to fly off helter skelter.

During her second winter at hack this buzzard was absent for a whole week, and I greatly feared that some pigeon-shooter might have found an all too easy target, even though she would no longer tolerate a close approach. There was much rejoicing when she again occupied the pine-tree perch. In March there were absences, but there was also much calling and another buzzard was frequently seen with ours. I concluded that food was being provided from this source. In the middle of the month she left us, presumably accompanied by her suitor. Perhaps it was a lull in family duties which enabled her to return for one well-rewarded visit on the first day of June after an absence of eleven weeks. I am optimistic enough to believe that one morning, glancing as usual at the now long empty perch, I shall see her sitting there as she did regularly for almost eighteen months.

Our next buzzard, a male, achieved a meteoric rise to fame in a few short weeks. Floundering in a ditch with a broken wing, he was taken into the care of Graham Moysey, who set the wing in plaster. It seemed a grim and unnatural business to keep a buzzard practically motionless for a whole week; but it was necessary to give the bones a chance to knit together. The treatment was successful beyond all expectations and, after a period of convalescence, this buzzard travelled to Bristol to be shown on television. A few days later he was taken on to the moor behind our house where it was intended to film his release. I had misgivings about this, since it was manifestly impossible to rehearse the event and I doubted whether the camera would be properly placed or the bird would even attempt to fly. However, with clear sunshine and scudding clouds the cameras had everything in their favour. A stiff breeze blew up the moorland slope on which we stood, buzzard on fist. The jesses were cut, and with outstretched wings he sprang into the air and rose easily skywards in fine style with a few powerful flaps. It was a thrilling moment, and I have never experienced so much excitement in watching a buzzard soar to a commanding height. Two crows attacked with shallow stoops, but with disdainful ease he mounted higher still, leaving them far below. He was joined by another buzzard, and together they circled till they seemed to be among the clouds high overhead. Eventually they parted and sailed away, drifting with the wind. We could only wonder at the astonishing transformation in a bird which had so recently been helplessly grounded.

The third buzzard, also a male, followed somewhat the same pattern. There was no broken wing, but he was picked up in a completely helpless condition under some wires into which he had presumably crashed. On arrival he was bleeding from the mouth and scarcely able to move. At first he spent all his time lying down, but he gradually recovered sufficiently to sit up normally on a perch. When he was later released, he flew off strongly and settled in a wood some distance away.

With his departure we could feel that there had been a happy release for all three buzzards. Yet they had been in such dire straits on arrival that a 'happy release' of an entirely different kind had at first seemed inevitable. We retained the phrase, referring to them as H. R. 1, 2 and 3. Any of them may be seen again. H. R. 3, the last arrival, was found not far off from where we released him; he could still be recognized by his jesses and unusually dark breast. H. R. 2, whose release was seen on television, was found some six miles away and may well be re-established in his home territory. Of course, we would be delighted to see our original buzzard, H. R. 1, back on the pine tree near the house, and we are still hopeful that one day she will be there.

Aggression at hand

Of four kinds of tit which were constant visitors during three winters to our garden near Bath, great tits were the boldest at taking food from us. The photographs (Illus. 1 and 2) were taken one day in February; the bird on the left landed first on my wife's hand, to be attacked by the pugnacious male on the right. He was very bold when taking nuts from us and spent much time clinging to the window-frames and tapping on the glass; early in the morning there was nothing for it but to get up and feed him. He had a bent tail, which my wife insisted was due to his habit of squashing it against the windows. Although no tits might be in sight, they would come straight to window or hand when we whistled. In really cold weather great tits have descended on me in the garden and pecked me in a demand for food. – *J. S. Beswick*

TWO TAME ONES by Lolita Alexander

Two mistle thrushes (Illus. 4) have been coming to my whistle for five years now, flying from anywhere in the garden or from across the road, where they nest in tall conifers. The first time the female landed direct on my hand she flew from the television aerial of a nearby house. The birds disappear for two or three months each summer after nesting and come back separately, tame as ever. One year the female brought two fledglings; while she fed one, she watched me feed the other, decided I was safe and left me to it. They finally fell asleep on the sill for about half an hour. Once or twice I had three birds on my hand. I started taming the female by throwing her currants and sultanas, progressively closer to my feet; then I substituted my hand, putting the currants first near the fingertips, later on the palm. She is still the tamer of the pair and now stands on one hand after eating the currants, ready to jump across for the cheese course. I have also had a mistle thrush on one hand, and on the other a young blackbird which became so tame that it would fight me to get into the house first. Taming the birds took much time and patience, but has been well worth it for the pleasure of seeing a large mistle thrush wheel in the sky and land, tail spread and wings out, with a thump on my hand.

JACKDAW JILL by W. Hacker

Hatched in May 1952, taken from the nest when half-grown, and hand-reared by my wife and myself, our pet jackdaw was well known for years as Jack, until her behaviour convinced us that Jill would be more appropriate. Most of her young life was spent in orchard or field, where she found many titbits uncovered by our hoes or under stones which she 'asked' us to turn over, revealing woodlice, centipedes and beetles. She was especially fond of cut-worms, which she pulled to pieces before eating, but she did not touch earthworms. On country walks Jill rode on my shoulder, from which she would spot insects at unbelievable distances, swoop to catch them and return to her perch. Nowadays she prefers cooked or uncooked pastry or meat.

When my wife went away for a few weeks in Jill's first September, the bird disappeared, returning the following February, half-plucked, dirty and hardly able to fly. She perched on the kitchen mantelpiece, put her head under her wing and seemed unable to take food. We kept her in an old parrot cage until she recovered, but she was left with the drooping wing by which we still recognize her. We supposed that other jackdaws had attacked her, as is usual when hand-reared animals try to join their wild relatives, although she returned to a free life, paying us regular visits almost daily. She peeps in until she spots someone who will open a window or door, then flies to a shoulder and expects to be fed. Years ago Jill found a bedroom window which was kept open, so she used to perch on the bedrail, waiting for us to wake. As soon as we spoke to her she would answer and expect a biscuit before flying off. She also joins us in the garden, sitting with eyes closed until addressed; then she opens one eye, hoping for her head to be scratched – against the lie of the feathers. As a young bird Jill was very talkative, but she hardly says anything now.

About six years ago she started coming for bread and milk seven or eight times a day, which decided us that she was feeding young. She has followed the same pattern since, and about four years ago she brought her mate, who was very shy and waited on the roof while she came inside. In time he began to come to the window-sill and eventually arrived on his own to load up, apparently when Jill was brooding the young. We have often watched her, after filling her pouch, take off towards the nearby village of Madingley, but we have failed to locate her nesting site. We have not seen her in a flock; she is either alone, or with her mate to whom she seems to be paired for life, or with her brood which, in the past three or four seasons, she has brought to the garden a few times. We have reared jays, magpies, rooks and kestrels; they all became very tame but did not return. Jill is the only faithful one, now in her fourteenth year.

Combined operation

A young house martin we found injured one August would feed only on moving insects; so I used to hunt for them with the bird perched on my finger. When it showed recognition of a settled housefly or blue bottle by a movement of the head, I would position it a few inches away, but there was no action until the insect moved. Then the martin's claws dug into my finger as its head darted rapidly forward, the beak engulfing the fly except for a few wildly thrashing legs, which soon disappeared. To give the bird a drink I dipped a finger in a bowl of water and placed a drop on its beak. If it was not thirsty, it gave a few twitches with its head; whereas the beak held skyward, with movements of the throat, indicated that the drink was appreciated. – *J. P. Gordon*

Voles on pipette

Rodents' milk is rich in protein, so Fenella Greig, a research zoologist, faced a difficult problem when she started to rear four deserted short-tailed voles about a week old. She decided to try undiluted cow's milk, which she carefully warmed to blood heat and sucked into a sterilized fine glass pipette; from this she could squirt it into the voles' tiny mouths (Illus. 5 and 6). At first she fed them every three hours, day and night. Taking each in turn, she held it in cotton-wool and allowed it to suck the pipette. We could watch the milk slipping along the tube into the gulping mouth. After ten days the voles began to lose interest in milk, so Fenella would sit one in the palm of a hand and hold a piece of carrot to its nose; placing a four-toed front paw on the carrot, it would start to gnaw vigorously. As wild short-tailed voles are fully weaned at fourteen to eighteen days, the family were now ready for their natural habitat of grassland or heather. Their weights had increased from about five grammes to fifteen grammes; at twenty grammes they would be in breeding condition. – *Andrew Pearson*

ALMOST NON-STOP by Stella M. Turk

We rescued a common shrew unharmed from one of our cats and put her in a glass-fronted cage with dishes of food and water. She quickly started to scrunch strawberry snails, dismember beetles and chew woodlice with rapid movements of her jaws and the air of one unable to get a meal down fast enough. Twice daily she ate a ration about equal to half her own weight of half an ounce.

It was a hot summer, and slugs, snails and worms were notably scarce. Although woodlice were abundant, our shrew was not partial to the commonest species, *Porcellio scaber*, perversely preferring the rarer and more slender *Philoscia muscorum*. She also showed a marked preference for the field slug, probably because this species has thin, and therefore not very sticky, slime. Before eating any slug she would rub and roll it vigorously on the floor with her long snout and tiny feet; and every few moments she would pause for a frenzied cleaning of her nose by drawing her forepaws repeatedly over it. This was a lengthy process with sticky subjects such as the great black slug, but she eventually demolished those nearly as large as herself. Some species of ground beetle put up a fight; the smaller were vanquished, but the much larger violet ground beetle, possessor of a most formidable pair of jaws, was taken from the cage unscathed next day. Harvestmen, despite their frail appearance, were not readily attacked, no doubt because of the fluid they are able to expel from abdominal glands; but they were readily accepted dead or injured from forceps.

Our shrew rested for only short periods. She would waken quickly at taps on the glass and come racing down the ladder from her sleeping-box, her long nose quivering. Because of the constant carnage and musky smell, we suspended her cage on an outside wall of the house, away from the reach of cats. She remained sleek and fat to the day of her death, cause unknown, about three months after capture.

TO TAME A VIXEN by Vivienne Letts

As evening approaches I walk into the road outside my garden and begin to call, a shrill cry that rises to a high note and falls to a low one. Josephine the vixen rarely comes first time; the call must be repeated. I have put a dish of food at the foot of a tree just outside her copse, but she is cautious at first, pauses, then makes an uneasy circling movement, sniffing the ground. Stopping again, she looks straight into my eyes, and we stare at each other in complete silence. She turns to the dish, eats the lot and retreats into the wood without looking back.

It all began one October day at dusk when I was walking through the woods with Cindy, my Shetland sheepdog. She ran off a little way, then came back barking. She had been following a vixen which travelled full circle and then, realizing that the dog was no longer chasing her, was coming up the ride towards us. This happened several times, and both animals seemed to be enjoying it. Eventually I caught the vixen's eyes in my torch and proceeded to talk to her in a very soft voice. She was obviously curious to find out what was behind the beam and kept trying to peer round it. So I turned the torch off. She did not run away on seeing my shadowy figure but sat down and listened to me and to the sounds of the evening. I decided to leave her and walk back the way I had come; Cindy, who is well trained, followed quietly.

Our next meeting with the vixen was accidental and in daylight. She ran off, so I spoke to her softly, using the same tone and phrases as before, and she turned to listen before going on her way. Once more we retraced our steps and left her in peace. I did not set out to tame Josephine, as we came to call her; it was like passing the time of day. But I felt this was a good opportunity to study foxes in the wild, and my husband and I started to go out every night with torch and binoculars. Time and again we bumped into Josephine and watched how she hunted up and down the road, from one side to the other. We decided to give her a little food in a paper bag to take back to her cubs. But we have not trained her to take food from the hand, and we beg others not to do this. After a few years an old vixen has to move to a new earth and, if she is hand-tamed, will only be shot.

We met Josephine when she was a year old and had her first litter, and she has been with us four years. We saw her mate, whom she had for three years, on several occasions and

could tell them apart by shape, colour and the white tips
to their tails. He was long-legged and very wild, but became
a little more used to us. While waiting by the copse, I would
sense something standing behind me and, turning round,
see his mask poking round the gatepost. I would talk to
him softly, and only then would he cross behind me and
over the road.

I have seen Josephine standing behind our two badgers in
the background of their track and a tabby cat mousing to
my right – all quite happy as long as they did not get too
close to each other. Several times the cat has been near the
vixen, who was interested in it; but it took no notice of her
and went back to its mousing.

I do not think Josephine had been tamed as a cub but
believe we were able to make friends with her by a combina-
tion of skill and quietness. The fixed friendly expression and
the right tone of voice are important; and the fact that we
did not follow her but walked away after each meeting gave
her confidence. Of course, we had the right circumstances,
living so near, and she was the right type. Perhaps families
of wild creatures often include an extrovert who is more
interested in people, as part of its environment, and less
frightened of them.

My husband and I have been studying foxes for seven
years. We have found them to be charming creatures and by
no means the rogues others would have us think them.

Part Three

Hunting and Feeding

Introduction

Almost all the smaller birds and mammals that can be watched in Britain spend much of their time looking for food. So it is inevitable that many of our observations concern their feeding habits, especially as we often directly or indirectly supply them with food. This Part is therefore composed entirely of notes, some long, some short, which cover a variety of animals and situations, starting with invertebrates: Stephen Dalton's beautifully 'captured' spitting spider, the tiny spiders living in a modern sewage plant and the curious anthomyid fly described for us by that doyen of field entomologists, Dr C. B. Williams.

The appetite of toads for bees seems well documented, that of hedgehogs is less well known and Doris Wright's note describes another death trap for these harmless animals, the cattle grid. Examples of the taste of adders for young warblers, of rats for snails and of foxes and long-eared bats for flies lead to notes illustrating aspects of birds' feeding habits: an ingenious starling, disciplined jackdaws. Winwood Reade's green woodpecker showed a remarkable sense of locality and several short notes testify to the feeding versatility of swallows, usually considered to be aerial specialists. Four more notes under the heading 'Opportunists' illustrate food-stealing by species that we regard as successful in the modern world; Robert Thompson's observation of the different openings presented to the starlings by the curlews and the oystercatchers is particularly neat.

Mark Taylor, an artist-ornithologist who is an annual tenant of the voluntary warden's hut on Havergate Island, describes the probable origin of the decoyman's little dog; and the Part ends with examples of how one mammal and several birds solve the problems of the feeding table. Dr Janet Kear of the Wildfowl Trust's research staff turns her attention to the finches, and F. Pitches shows how these problems can be elaborated into intelligence tests.

THE SPITTING SPIDER by Stephen Dalton

Slowly making its way towards a mosquito on the window-pane is an odd-looking light-coloured spider. Its long front legs grope from side to side, as though it were blind and feeling its way, until they touch the mosquito, which remains undisturbed on the glass (Illus. 21). If you now look at the spider closely, you will see that it rears up on its 'haunches' and jerks its body. Suddenly the mosquito starts to struggle, as though its feet are pegged to the ground. Through a magnifying glass you will see that it is indeed held down, rather like Gulliver, with a series of sticky-looking threads which lie, zigzag fashion, on both sides of its body.

The extraordinary creature aptly named the spitting spider (*Scytodes thoracica*) is one of the most fascinating British species; yet most people are unaware of its existence. This is hardly surprising when one considers that its body is a mere six millimetres long, and it comes out only at night to hunt its prey. In England it was first recorded by W. E. Leach in 1816, and in more than a hundred years only six other specimens are known to have been found. Recently, however, it has been seen in most counties in the southern half of the country, and in some areas it has even been found quite commonly.

I kept the specimen illustrated in captivity during most of one summer and often watched it take its prey. As Dr W. S. Bristowe has explained, the spitting is accomplished so quickly that the eye cannot follow it; but during the split second when the spider jerks its body, the jaws appear to oscillate from side to side as the spit is discharged. The unusually large front half of the creature's body is mostly composed of huge glands containing gum and poison, both of which are connected with the fangs. If the mosquito persists in struggling, the spider bites it in a leg and it soon dies from the effects of the poison. The spider then pumps digestive juices into its prey, so that the body contents become partly digested and fluid, and the liquid can be sucked until nothing but the empty carcass remains.

GROUND SHEET by J. A. L. Cooke

The filter beds of modern sewage disposal works have retaining walls of stone or brick, which enclose a bed of pebble or clinker some four to six feet deep. This is watered every few minutes by partially purified sewage from a rotating arm. In these moist and unwholesome conditions a rich growth of algae, bacteria, fungi and protozoa soon develops on the surface of the clinker; and within this slimy gelatinous covering is a habitat highly attractive to worms, flies and various other small creatures. Indeed, certain flies, notably a chironomid midge and two psychodids, may breed in such abundance that they can pose a serious hazard to people living nearby.

The natural controls of this potential menace are small unassuming black spiders no larger than a match head. Studies on them are not numerous, but of the fifteen species recorded, three at least (*Lessertia dentichelis*, *Erigone arctica* and *Porrhomma convexum*) appear to be specialists in combating the insect squatters. They all belong to the family *Linyphiidae*, better known as 'money spiders', which make up about a third of the British spider fauna of six hundred species. Most of the year they live deep in the filter beds, spinning tiny webs between the pebbles and clinker, but on fine autumn days they tend, like all money spiders, to make aerial dispersal flights, by extruding fine strands of silk into up-draughts of warm air until they are themselves wafted skywards. Thus they can journey hundreds of miles, crossing oceans and continents without difficulty.

If the urge to migrate occurs when there are no suitable currents, the spiders run hither and thither, trailing silk behind them. But, instead of being carried away, it lies on the ground and soon builds up into a continuous sheet. The photograph by Andrew T. Paton (Illus. 10) shows a filter bed in West Suffolk which has become clothed in a shining film of white silk. Occasionally such sheets are carried skywards in quantity, to float away as gossamer.

FUNGUS-EATING FLY by C. B. Williams, FRS

In late spring and early summer, especially in damp shady localities, one sometimes sees a whitish fungus encircling the stems of grasses in the form of a cylinder an inch or two long. The grass is often cocksfoot, though many others, including couchgrass, bent, foxtail, fescue, Yorkshire fog and timothy, are also attacked. The fungus, *Epichloe typhina*, is known as either the choke or the reed-mace fungus, a name derived from the development rather later of reddish-brown fruiting bodies, so that the whole somewhat resembles a miniature head of the common reed-mace. It can be a minor pest of grass seed crops by preventing seeding, the portion of the plant above the point of attack often being distorted.

A close examination of several of these fungus growths is likely sooner or later to reveal the tiny boat-shaped elongated egg of a small two-winged fly of the family Anthomyidae, known as *Chortophila spreta*. The egg hatches on the underside, and the small legless maggot feeds on the fungus, making furrows or tunnels in all directions; when not feeding, it usually returns beneath the eggshell. The fully fed larva falls to the ground and forms its puparium, in which it remains until the following spring.

Some years ago, when I was at the University of Minnesota, I found an apparently similar fungus on grass in the neighbourhood of St Paul, and on it were eggs and feeding marks exactly similar to those of our *C. spreta*. I have little doubt that the fungus and fly are identical with ours, but neither had been previously recorded in the USA. Even before this, on the island of Trinidad, I had found a very similar egg of a small two-winged fly whose maggot had the same feeding habits; it was living on a fungus that was parasitic not on a grass but on another insect, a rather large species of leaf-hopper. Thus we had a fungus parasitic on an insect, and another insect eating the fungus – an excellent example of 'big fleas' and 'little fleas'. Although the fly on the reed-mace fungus is apparently widely distributed, the exact limits of its occurrence in Britain are not at all well known.

Toads and bees

We noticed a small toad, about two inches long and greyish-black in colour, sitting in front of one of our beehives, perfectly placed to catch any bee coming out; its lower lip was just on the alighting-board. I picked the creature up and felt its bread-basket, which seemed quite full of bees. Although we knew toads to be great insect-eaters – their capacity is enormous – we hated to kill it; so we carried it some three hundred yards to a large patch of underbrush. The next morning it was back in position at the hive entrance. We knew it was the same toad because it had no right eye. This is the second time in fifty years of bee-keeping in Florida that we have lost a swarm to a toad. – *J. H. Tullis*

While on holiday near Dunkeld, Perthshire, some years ago, on a fine muggy evening after a wet summer's day, I accompanied my host to see if any toads were 'at the bees'. Sure enough, on the alighting-board of a hive a toad was flicking into its mouth every bee that tried to enter. – *Thomas Watson*

During the war a swarm of bees took up residence in a hole in the mound of earth covering our air-raid shelter. One evening I found the entrance to their nest blocked by a sleepy and apparently well-fed toad. Not a bee was to be seen, and they never returned. – *R. W. Brown*

Herrod Hempsall, in his *Bee-keeping New and Old* (1937), lists the toad as a pest; but some may eat chiefly the old or diseased bees and so do more good than harm. – *S. E. Alsop*

For ten minutes I watched a toad sitting by a wasps' nest in a garden bank flick its tongue in and out, capturing a wasp each time. – *W. H. Tarn*

BEE FEAST by Doris Wright

Knowing the unpleasant fate that awaits hedgehogs which fall through cattle grids, I regularly inspect all near our farm at intervals of a few days. Rescue is difficult, as I have to nip one prickle between finger and thumb and pull the awkward body up to and through the bars. One summer morning I freed one and slipped it over the garden wall. About twenty-five yards off was a beehive under which a swarm had formed, and I was on my way to meet a friend and deal with it. When we reached the hive, we found that the hedgehog had beaten us to it and was running straight into the swarm until its head was in the middle of the moving mass of bees. Then it backed out munching, swallowed and waded in again. This in-and-out action, followed by a rapid chew and swallow, continued for several minutes until the starving animal had taken the edge off its appetite. It then picked up a few more bees from the outside of the swarm, turned and ambled off into the bushes. I noticed what appeared to be dozens of tiny nylon spikes in its snout; they may have been its own hairs standing up on a very swollen nose, though it is hard to believe that the bees would not have left their traces. [The thought of how hungry the hedgehog must have been to try for such a dinner may persuade readers to inspect cattle grids and, if possible, put in a plank up which stranded animals can climb to safety. – *B.C.*]

In the raw

The alarm calls of a pair of willow warblers attracted my husband's and my attention as we were walking in the New Forest in June. They were flying backwards and forwards, low over the ground. Seeing a movement in the grass, we approached to within two yards of an adder swallowing a nestling. This was still alive in the snake's jaws, but after a few seconds it stopped struggling. A bulge farther along the sinuous body suggested the fate of another of the brood. For several minutes we watched the snake dragging its prey away, while the parents continued to dive at it, landing less than a foot from it and completely ignoring us. Unfortunately the snake must then have seen us, for it moved suddenly into dense cover, dropping the now dead nestling, and coiled up at the base of a small hawthorn bush. There we left it, with the warblers still calling persistently and attacking it. – *Stephanie J. Tyler*

The gourmets

A low stone wall separating two flowerbeds at different levels is the haunt of snails, which in winter remain tucked in the crevices. Having several times noticed a large brown rat scuttling across the garden, I decided to watch it one wet and blustery day. It came out of a hole hitherto masked by nasturtiums near our fuel bunker and, after a quick look round, crossed the path to the bed below the wall. Here it stood on its hind legs to peer between the stones, turning sideways to expose its pale belly as it tried to scrabble into the narrow gap. Foiled, it began to dig in the soft bed, scraping with its forepaws and throwing earth out between its hind legs; every now and then it would glance behind it.

Then it got its head well into a gap, emerging with something between its jaws. Returning to its hole, it reappeared just beyond the far wall of the bunker, which is my boundary. It popped out of a corresponding hole in the next-door yard, before vanishing into a drainpipe; and a moment later it retraced its steps to the wall for another spell of digging and scratching. The prize was a snail shell, and it must have taken away at least a dozen while I watched. I went round into the yard but, though the tunnel entrance was easy to see, I failed to locate the store-room. Some weeks later I came on a heap of empty unbroken snail shells, but I have no direct proof that the rat had collected them. If it did, then it was able to extract the snail without breaking the shell. – *Margaret B. Kabell*

The rockery in our back garden used to be the haunt of a colony of snails; but one hard winter a brown rat took a fancy to them. Several times I watched it lever them with its front paws from under a particular stone, one after the other, until the store was depleted. With a shell in its mouth, it would crack this against a rock, eating the contents immediately. – *Dorothy Spearing*

Wormed out

By mid-summer our Surrey lawn had reached a state of near-perfection. But early one morning I saw that a mole had been burrowing along one edge; for about fifteen feet a succession of mounds curved in half-circles, then turned back like an 'S' on another track, with here and there an exploratory mound diverging from the main line. Later one of my grandsons said he could see movement in a side tunnel. We watched the soft earth being thrown up, then to our amazement a large earthworm came out of the loosened end and moved at great speed straight across the grass for about ten feet before entering the earth again. As soon as it emerged, a mole's head appeared from the tunnel and remained at the opening for several seconds; but it made no attempt to follow the worm. It soon drew back into the hole, and the burrowing ceased. – *S. Winifred Harper*

Fly-catching fox

One September day a gamekeeper friend watched through binoculars a fox in a field about a hundred yards away. It performed what looked like some strange ritual dance: crouching, rising in the air and pouncing. Eventually our friend realized that leather-jackets were hatching out into crane-flies, which the fox was hunting and eating. In a quarter of an hour, he estimated it accounted for forty to fifty of them. Foxes will also nose out the insects which breed under cowpats and other animal droppings, to provide a feast of beetles and grubs later on. – *Basil Cullin*

Off the glass

About midnight on October 1st eight or nine large moths were going up and down the panes outside the window of my bedroom, in which the light had been left on. Suddenly one vanished as if it had never existed, then a second and a third. When the fourth disappeared I heard a faint click as something touched the glass; but a moment or two passed before I realized that a bat was feeding off this heaven-sent horde. I could not have had a better view as the animal flew up to the window at speed, nipped a moth head on to the glass and turned smartly to its left without touching the glass again. When there were only two moths left, still vainly trying to reach the light, I switched it off and crept into bed in the dark. I could not be certain of the species of bat, but the date, time of night and method of feeding right off the glass suggest a long-eared. – *W. T. G. Boul*

Food in comfort

A pair of wrens with their four fledglings haunted the gnarled plum tree in our Yorkshire garden. The flat stump of a sawn-off lower limb made a splendid bird table but the wrens, darting among the branches above, left the food alone. Nothing we put out tempted them, and for a fortnight they tantalized us with their quicksilver movements and shrill calls resembling the squeaking of an uncoiled hinge. Then one evening the four young ones emerged from cover and with one accord flopped into the still perfect cup of a blackbird's nest on a low branch immediately behind the bird table. They fluffed and wriggled themselves into comfortable positions, lifted their heads and opened their beaks wide. A moment later the parents touched down with beaks full of food which they stuffed into each yawning mouth. The activity went on for almost ten minutes before all six flew away. We did not see them again. – *Joyce Fussey*

Using the wind

A starling was trying to 'hook on' to a spiral food basket hung from an eastward-projecting branch which was swaying in a fresh westerly wind. Each time that the bird took off downwind from the tree, it was carried rapidly past its objective. After a number of attempts it retired to the tree. Then, to my surprise, it took off into the wind and let itself be carried slowly backwards to the level of the spiral. After one or two trial runs, a quick grab enabled it to get a mouthful of fat before being blown past. – *R. T. Shepherd*

Your turn, Jack

The peace of a Sunday morning was shattered as a throng of jackdaws fought over some stale bilberry tart on my bird table. Suddenly there was silence, and from a window

I saw a single bird on the table, the rest squatting on the grass below. When the 'king-pin' had finished eating, it flew to a nearby branch and sat cleaning its bill on the bark. The noise did not break out again and the tart was taken over by a second bird, the rest still waiting in silence. By the time the two or three tail-enders had shared the last morsels, the first bird had flown away, but there had been no more squabbling. It seemed a perfect example of the pecking order in practice. – *Sheila Harries*

Food in winter

The raised rockery in our Oxfordshire garden measures thirty-six feet by four feet, but January's snow drifted till it became a featureless sheet of white continuous with the lawn. Two days later a green woodpecker pitched in a birch twelve feet away, then flew diagonally down to the rockery. It immediately started turning round and round in the soft snow, wings held slightly out, as though making a nest-scrape or a bowl with its body. Having thus lowered the level by about four inches, it began to sink a shaft with its bill. It virtually cut a section, slicing in half a clump of saxifrage below which there is usually an ants' nest. The finished shaft was eight inches deep; there were no false starts or trial borings, and the snow round it was untouched. Somehow the woodpecker had visualized the position of its customary source of food, although it lay buried under an expanse of white. It was not until I saw the saxifrage emerge that I remembered the ants' nest there and appreciated the remarkable precision of the whole operation. – *Winwood Reade*

Several times in the winter of 1962–3 I drove one hundred and seventy miles through the 'corridor' between Berlin and Hanover to see my wife, who was in hospital there. On five separate weekends I travelled through a blinding snowstorm; the last time the snow lay feet deep at the road-side, and the flat country around, always desolate, was without any relief from the vast carpet of whiteness. At Marienborn I pulled into a lay-by for a drink of coffee. As soon as the car came to a stop, two blue tits flitted from the trees, landed without hesitation on the arm of the windscreen-wiper and began to tap at the glass. I hurriedly unpacked a sandwich and opened the window so that I could sprinkle bread on the bonnet. The tits flew back into the trees until I had closed the window again, but returned immediately and drummed away on the bonnet until most of the crumbs were gone. – *Arnold Bosworth*

It started when someone threw bread to the ducks standing on the frozen surface of a Cheshire pool. They hurried towards the food and found that walking on ice is not as easy as swimming in water. First one foot would shoot out sideways, then the other; or both would shoot out at the same time, so that the birds landed on their stomachs. When eventually they reached the bread and stretched out with their beaks to snap it up, it shot across the ice as soon as they touched it. Persevering, they slithered off in pursuit. Sometimes one would intercept another's bread, reminding us of a game of ice hockey as each tried to gain the prize for itself. Appropriately this pantomime took place on Boxing Day. – *Nancy M. Kelcey*

Adaptable swallows

For many years I have fed birds all the year round on a bird table attached to the kitchen window-sill, and also on the ground below. On June 4th a swallow alighted on the grass and for two or three minutes ate brown breadcrumbs with the sparrows, running to pick up pieces and break off small portions. – *Heather K. Rudkin*

About a hundred swallows were wheeling and circling over my garden one morning last August; then some thirty or forty settled on the roof of the house, which is Roman-tiled, and stayed there for fifteen minutes, flitting about and pushing their heads under the ends of the tiles. Later we found that a swarm of flying ants had landed on the roof; evidently the swallows, after pursuing them in the air, had followed them down. This was the first time I had seen swallows take their prey other than on the wing. – *M. E. Pearce*. [Although swallows are known sometimes to pick insects off roofs, trees and the ground, these two observations are most unusual, particularly the bread-eating, since the swallow family is regarded as insectivorous. – *B. C.*]

Three times in July and August I watched swallows hawking where the River Almond enters the Firth of Forth. Most of them remained on the wing, but several landed and, after resting, spent up to ten minutes taking live food from among the shingle and boulders, keeping strictly to the high-tide line where insect life is most prolific. I believe they were catching *Calopa frigida*; there was no obvious hatch of another fly. Midges near trees only fifty yards inland were ignored. – *Patricia Veall*

Early in August in Aberdour harbour, from a distance of twelve feet, my daughter and I watched a solitary swallow pecking for about ten minutes among the stones at one spot on the shore. It was a fine warm evening with the tide coming in. – *Emily Manini*. [Aberdour is on the opposite side of the Firth of Forth to the mouth of the Almond where Patricia Veall watched her swallows. – *B. C.*]

We were bird-watching from the car near the Zwartkops estuary, Port Elizabeth, when five ruffs landed nearby and began to remove something from short bushy plants growing on the roadside. It was a cold overcast day with a moderate to fresh south-east wind. Soon they were joined by up to twenty European swallows and a few South African sand martins, which kept flying to and fro over and round them, apparently catching insects disturbed by the ruffs. A gnat which I identified as *Polypedilum lobiferum* must have been present in millions. Swallows often hawk insects disturbed by large animals walking through fields or veld; but this is the only time I have seen them 'cash in' on other birds. – *J. S. Taylor*

Cyrano

While we were feeding waterfowl by a lake, a solitary cock house sparrow flew down. He had a malformed beak, the upper mandible being twice as long as normal, flat and shaped like a duck's bill. We were wondering how he could feed himself when he turned his head until one cheek and the monstrous mandible were resting on the ground, then picked up a crumb with a sideways movement. He had to have several tries at each before getting it into his mouth, and had not finished his meal when a hen sparrow appeared and carried off the choicest fragments. It was clear why Cyrano had to forage on his own. – *Freda Hurt*

GREAT GREY SHRIKE by H. R. Tutt

Although it is a regular autumn and winter visitor, the great grey shrike is rare enough at any time to cause excitement among bird-watchers. On October 18th, 1960, one was seen a little before noon on Havengore, the most westerly of the level marsh islands that lie east of Shoeburyness. In mid-afternoon another observer spotted it at Foulness, skimming along the sea wall and out on to the saltings. He called a friend and, as together they went over the wall to look for it, two small birds flew overhead, uttering a call note with which one of the men had become familiar from studying four bearded tits that appeared near Southend-on-Sea in 1959, when the great scatter of this species from Minsmere occurred. Strange it seemed to see two such local rarities at the same place on one day. Eventually the men spotted the shrike again and, standing under the sea wall, watched it swoop on one and then another of several meadow pipits that were on the wing; each escaped by side-slipping as the aggressor descended on it. The shrike then flew to an old stake, split at the top, and took from it a headless bird, which it carried to barbed wire. After sticking the prey on a barb, it proceeded to tear it to pieces, which it devoured. Later a tail, two covert feathers and a foot picked out of the herbage confirmed that the victim was a bearded tit. Unfortunately the shrike was not seen capturing the tit. In his book *Birds of the British Isles* Dr Bannerman writes that he has never himself seen a great grey shrike attack and kill a bird, nor has he read an account of how this feat is accomplished. In Canada in 1957 I visited the log cabin, far out in the bush in Ontario, of a noted bird observer, Mrs L. Lawrence. She related how she had seen a northern [great grey] shrike swoop on a black-capped chickadee [willow tit] and seize it in the claws of one foot; she added that, if she had not seen it herself, she would not have believed it possible. [In Scandinavia 1960 was a 'lemming year', and large clutches are reported for great grey shrikes, one of the lemming's predators. This may account for the exceptional number seen here in the autumn, when several reached Ireland. – *B. C.*]

Dance of death

The sparrowhawks have returned to our fellside. Neighbours reported them swooping into the garden and carrying off a juvenile blackbird and cuckoo. A tame young robin somehow escaped the clutching talons, fluttered into a dense thicket and was not seen again. But it was the noise at the first greying of dawn that impressed us most. Both households had sparrow families nesting under the eaves and heard the commotion: a banging and dancing on the roofs that woke us up and made further sleep impossible. I stole outside to see four sparrowhawks apparently carrying out a war dance on the roof directly above the nesting sparrows. The half-light was confusing, and I could make out little of what went on; my neighbours were more fortunate and saw the hawks in clearer dawn light. The birds were capering along the edge of the barge-boards, fluttering their wings in readiness to pounce; and, though they could not be absolutely positive, my neighbours were almost sure that the banging was made by their beaks. The hawks seemed to know what they were after and how to get it, for no sparrow chirps under our eaves now. – *Anne Montrose*

Winner lose all

Three starlings and two great black-backed gulls were feeding on a Northumberland bowling green when a race for a tasty morsel developed between one starling and a gull. The smaller bird swallowed the prize, whereupon the gull moved forward swiftly and smashed the skull of its rival, whom it swallowed whole; the legs were still kicking as they disappeared down its gullet. The other two starlings flew off at once, screeching. The gull, finding no more morsels, also departed. – *H. A. Molloy*

Opportunists

A cock blackbird, foraging on the lawn, had started to dig up the grub of a large beetle, when a cock house sparrow flew down and took up a position five or six feet to the rear. He watched the digging, which became fast and furious as the blackbird's beak reached the quarry. Suddenly the digger gave a mighty heave and out it came. The blackbird failed to hold it and the grub sailed through the air, landing within a couple of feet of the waiting sparrow. He was on it in a flash and, though it was almost too heavy for him, managed to fly off with it to a tree, to the blackbird's obvious dismay. – *A. B. Phillips*

On the lawn a song thrush was dealing with a titbit from the bird table when a cock house sparrow approached in a rather aggressive manner. The thrush opened its beak to threaten the intruder and dropped the titbit. The sparrow nipped in like a flash, picked it up and flew off triumphantly. The thrush stood with its beak open for about two seconds, as if flabbergasted. – *H. Batterbee*

Parties of curlews and oystercatchers, in roughly equal numbers, were feeding on the Maes-du golf links at Llandudno early in September. One morning between 8 and 9 AM I noticed a starling among them and soon saw it dart forward to snatch a worm just as a curlew with its long bill succeeded in drawing one out of the ground. The starling repeated the performance, beginning its predatory flight from a yard or two away and robbing more than one curlew. They seemed to take no notice and continued to feed at high speed. After a few minutes a second starling arrived and behaved in the same manner until the flock was disturbed. Only curlews were robbed, and I noted that it took them longer than the oystercatchers to transfer a worm from curved bill to throat, thus giving the starling more time in which to snatch it. It is also possible that oystercatchers would retaliate. The victimized birds may have been juveniles, but I could not distinguish them from the other curlews. – *R. M. Thompson*

Strolling on a deserted Cornish beach early one morning, I saw a herring gull swoop down and rise from the water holding in its beak an eel, which it dropped on the sand. Then it repeatedly backed a few feet, ran forward and pecked the eel. Watching from a nearby sandbank was a jackdaw, which presently flew down, alighted gently behind the gull and gripped the tip of its wing as it advanced to peck the eel. The gull at first appeared oblivious of the jackdaw, which eventually lost its grip; and the same thing happened on the gull's next advance. The jackdaw watched another attack without moving, then seized the gull by its tail and this time sustained the hold with great determination, all anchors down. The gull, now forced to give ground, was hauled backwards inch by inch. Suddenly the jackdaw let go, darted forward and in a flash was flying away, the squirming eel dangling from its beak. – *Joyce Hardiman*

DANGER AT A DISTANCE by Mark Taylor

In the upper Cherwell valley the early spring floods had stood high, covering the low-lying meadows between river and canal. They had provided safe feeding grounds for flocks of wigeon, mallard and teal, with occasional less common visitors to our narrow waters such as shoveler, pochard, tufted and pintail ducks. At the end of March, when I was walking the valley, the river was nearly back to normal and there remained only one swampy area large enough for the duck to gather safely. Small wisps of snipe rose at intervals from the reeds beside the canal, one or two curlew were trilling and the piping of redshank could be heard from the ditches.

At a distance the duck-marsh showed unusual activity, and I crept down under cover of a line of pollard willows. From the river bank I could see across the meadows to a rough thorn hedge some three hundred yards away on the other side of the valley, dividing the level ground from rising arable land. Patrolling in leisurely fashion among the dead grasses at the foot of the hedge was a fox, which stopped now and then apparently to listen for the stirring of mouse or frog. Twenty or thirty yards on my side of the hedge, abreast of the fox, were some forty mallard and twenty teal, keeping up with him as he advanced. It was hard walking even for the mallard, who had to waddle energetically to keep their places; stragglers to the rear would make short flights over the heads of the others to regain the lead. The teal were generally farther from the fox on ground a few inches lower, so that their heads looked like little black crotchets bobbing about in the rushes against the light of the low sun glinting on the meadow. One pair of pintail in the line of the mallard's advance let the crowd pass on either side of them, sitting with necks at full stretch and bills almost vertical to keep an eye on the danger. A standing curlew and heron also made no visible movement except when the heron caught a frog among, and probably flushed by, the duck procession. For nearly two hundred yards the fox strolled on, paying no attention to his followers, who were always too far away to be caught before they could take wing. Then he left the valley through a stack-yard, and the crowd dispersed within a few minutes to feed and preen, while the pintail lowered their heads and went to sleep. It seems to be accepted that just such a sight as this gave our ancestors the idea for the carefully constructed duck-decoy with its trained fox-coloured decoy dog, but the reactions of ducks to a fox in the wild are not often recorded.

Stoat at table

One cold March day I saw a stoat on my Perthshire bird table; it was in almost full winter coat of ermine, with brown patches only on the face and shoulders. It tore off a piece of suet about the size of a hazel nut, neatly jumped the two-and-a-half-foot gap from table to post-and-rail fence and ran along this to a drystone dyke, where it popped into a hole. Shortly afterwards it reappeared and twisted, leapt and somersaulted on the wall before returning to the table for a second helping. The birds left the stoat to dine alone, but within a few minutes of its final withdrawal they came back, seemingly quite unperturbed. – *S. R. Davidson*

High IQ?

One winter, on a tree in our suburban Bristol garden, we hung a wire spiral filled with peanuts for the birds. The grey squirrels soon spotted it but, because it was suspended on quite a long wire, they had great difficulty in reaching it; and the narrow diameter prevented them from extracting more than the occasional nut from the top. Only one solved the problem. Hanging by its hind feet from the branch and grasping the spiral with its front paws, it lifted this sufficiently to unhook it and carry it off like an oversize shopping basket. But when the method misfired and the container fell to the ground, intelligence seemed to fail; the thief's efforts to bite through the wire were unsuccessful, and it had to be content with the nuts which had shaken out. – *Richard Tinklin*

When I suspended half a coconut on thin wire about one foot below the branch of a tree, a grey squirrel tried somewhat futilely to slide its front paws down the wire, keeping hold of the branch with its hind paws; but the coconut tended to swing and the animal made no progress towards the 'meat' in the shell. It then sat up on the bough, leant downwards and with a front paw started a pendulum movement of the wire. When the coconut was moving through a sufficiently wide arc, the squirrel gave a sharp push and swung it into a fork of the tree, where it ate the contents in comfort. – *J. W. Gittins*

ADAPTABLE FINCHES by Janet Kear

We often find that birds of different species but with the same ecological requirements cannot share a habitat. Food is probably the most important limiting factor, and the continued existence of two related species in one place will usually depend on differences in their methods of selecting food.

Three years ago a paragraph appeared in *The Countryman* on the subject of adaptable finches which had learnt to feed like tits at bird tables; and the letters that followed dealt entirely with chaffinches and greenfinches. These two species will live happily side by side, although there is an overlap in the size and nature of the seed they will select in captivity, and both are known to be mainly ground-feeders. So other differences in feeding technique, such as their ability to feed at suspended baits, were worth investigation.

Of the two letters describing the habit among chaffinches, one mentioned the bird's sliding down string to perch on fat in the manner of a tit, undoubtedly unusual for a chaffinch; and Dr W. H. Thorpe has written of two further instances of these birds clinging to thread or fat. Seven letters recorded chaffinches solving the problem in another way, by swooping at the bait or hovering by it and feeding on the wing. Of nine records of the greenfinch, two described the bird in an upright but presumably unstable position on swinging bait; three others slid down string to reach the food. Six were observed to feed upside down, and Dr Thorpe saw three feeding together in this way.

H. Boase has written of individual greenfinches that once or twice fed on suet hung up for tits. He also noticed that, when chaffinches made raids on fat, they did so by hovering and pecking. Thus, although these two species seldom feed upside down, the greenfinch is apparently able to do so more easily than the chaffinch, which hovers at food with some frequency.

It may seem that competition between the species is not thereby reduced, because at certain places each can obtain the same food by a different method. However, after many hours of observation in the wild it was obvious to me that the greenfinch used its slight ability to hang while feeding at twigs and fine branches which the chaffinch never reached; and the latter employed its hovering technique mainly for catching flying insects. So these feeding postures can be related to the type of food that the birds obtain and will tend to spread their feeding activities through the habitat, allowing each species to exploit different food sources within it.

Reasoning crows?

I hang suet on thin strips about ten inches long from the cross-bars of our clothes-posts for the tits. It is amusing to see crows, after eyeing the morsels from the ground, fly up on to one of the posts and make clumsy and vain attempts to reach them. Last winter I watched one crow stand on a cross-bar and, after a few minutes' apparent study, reach down as far as it could, grasp the string with its beak and haul it up length by length till the suet was at bar level. As the bird was swallowing a piece it would accidentally let the rest fall, and it hauled the lump up five times. Surely this is evidence that at least some individual birds possess the power to reason? I have not seen a crow repeat the operation since. – *F. C. W. Stevenson*

One summer afternoon a crow, with a second bird in attendance, landed high on a tree in my neighbour's garden and gradually hopped down to a lower branch, from which a piece of meat was hanging by a length of string. The bird's claws were on the knot holding the string, and it tried several times to reach down to the meat; but it found this to be too far away and had to use its wings to recover balance. After a pause the crow bent down as far as possible, took the string in its beak, pulled it up and stepped on it, repeating the procedure until the meat was drawn up to the branch and it could make a meal. – *Frank C. Edwards*

Some years ago I made a bird table with a thatched roof, from the ridge of which a food basket hung; and to keep out the rooks I put bars all round between the roof and table. For a time they were baffled until one found a solution which others soon followed: with its bill it could just reach the basket, which it pulled against a bar and held there with one foot while it got at the food. – *C. J. Jacobs*. [In *Learning and Instinct in Animals*, W. H. Thorpe refers to a record of a captive crow in Germany. But there are few reports of the intelligent crow family adopting a method which seems to be instinctive to several small finches and is often mastered by tits. – *B. C.*]

INTELLIGENCE TEST FOR SPARROWS
by F. Pitches

Inspired by a series of intelligence tests on tits carried out by M. Brooks-King and H. G. Hurrell, I devised an apparatus for a similar experiment with house sparrows. As can be seen from the photographs (Illus. 11), matches are inserted through holes in the perspex front of a vertical wooden tube. A small transparent plastic tube containing bird seed is inserted into the top, its progress downwards being arrested by each match in turn as the one above is removed.

When all the matches have been extracted the seed drops to the level of the tray and becomes available to the bird. I set up the apparatus near a window with, at first, only the bottom match in position, and sprinkled a little seed on the tray.

One morning about a month later I noticed that the match had been removed and the seed container emptied. It was soon evident that a hen sparrow was responsible. She quickly became a frequent visitor, unhesitatingly removing the match whenever she found the tray empty. Over a period of a week I gradually increased the number of matches to six, and the same bird quickly mastered the technique of starting at the top and removing all in turn, before dropping to the tray to eat the seed. She also became tame enough to allow any of my family to watch her fascinating performance from quite near the window. Among the many sparrows which frequent our garden, only this hen has learnt to associate the removal of the matches with the appearance of the seed. Even after she had mated and raised a family in a nesting-box under the eaves, the one surviving offspring never learned the trick, though it perched on the apparatus many times and watched its mother at work. Other sparrows, too, come to the tray, and it is amusing to watch them greedily squabbling over the seed after their more talented companion has removed the matches.

Part Four

Confrontations

Introduction

Most fights in the wild have to do with food or mating. Most of the notes in this Part could therefore appear equally well in Part Three or Part Five or even Part Six; but I have tried to assemble here observations in which the fight itself, rather than its cause or effect, is the point of greatest interest: all of them refer to vertebrate animals, and most of them to birds, mammals, or birds versus mammals.

Some, like the antics of the hoodie and the rabbit watched by the late Stewart Campbell – whose cromag, fashioned by a third Campbell, I still treasure – and of the rabbit *vis-à-vis* the pigeon, as seen by J. T. McLean, are baffling. This second rabbit seemed to be executing an inefficient copy of the dance used by stoats to mesmerize its own kind. But D. Knowlton's rabbit turned the tables on the ancient foe, like James Ker Cowan's pygmy shrew and the cat.

Stoats and weasels figure prominently in these notes, because they are sufficiently rarely seen to make us stop whenever we have the chance to observe one, and because they seem remarkably indifferent to us when really on the hunt or at the kill. Raymond Hewson switches from his favourite mountain hares to describe the stoat's methods, and so leads to a group of notes which dramatically support his suggestion that kestrels (and other birds) may follow stoats and weasels in the hopes of an interception.

1. A great tit lands, followed by the male on the right

2. He claims and wins the nut by superior aggression

3. Heron in a Wexford garden

4. Lolita Alexander with two mistle thrushes

5. Vole on pipette

6. Feeding on a nut held in the hand

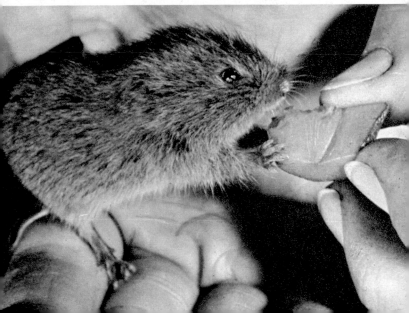

7. Stone curlew with newly hatched chick peeping from under left wing

8. After half an hour in a thunderstorm

9. Birch-leaf shieldbug guarding her eggs

10. Sheet of white silk on filter bed

11. (*left to right*) Hen house sparro
until see

removing and dropping matches
appears on tray

12. (*above*) Hen pied wagtail posturing; 13. (*centre*) flinging herself
against mirror; 14. (*below*) gazing from top

15. Jay in full anting posture

16. Starlings sunbathing

17. Newly born mole

18. Badger cubs on Somerset farm

19. Lapwing on nest in snow

20. Red-backed shrike at nest

21. Spitting spider feeding on mosquito

Slow-worm duel

On May 2nd, a gloriously sunny day, my husband and I were
walking in a Somerset lane when we saw what appeared to
be a long slim snake. On looking closer, I realized it was
three slow-worms. Two were lively and holding gently in
their mouths a third, which lay still. One on each side, the
active pair were pushing and pulling, apparently trying to
get it nearer the bank, so I thought it had been injured. One
then let go, whereupon the second leapt at it, coiling and
springing in the air, and caught hold with its jaws in a tight
grip just below the head. The two creatures twisted and
twirled and tied themselves in knots. The throat hold was
changed, and they locked themselves together, mouth to
mouth, still fighting. Meanwhile the third slow-worm came
very much to life, made for the bank and was soon out of
sight. Thinking there was going to be a fight to the death,
my husband parted the contestants and put one farther
down the road. The other then made straight for the bank
where the third had disappeared, and we realized that we
had seen a duel fought for a slim glistening lady. – *Violet I.
Ricketts*

Blown up for safety

On a garden path at the foot of a drystone wall a large grass
snake lay with the head and shoulders of a toad firmly
embedded in its jaws; obviously deadlock had been reached.
Carefully, and with difficulty, I disengaged the toad which
had a wound across its throat and chest. The snake disappeared into the wall but the toad, which had inflated itself
to avoid being swallowed, lay there and gradually subsided
to its normal size. Five minutes later I came out of the
house and, on seeing me or hearing my footsteps, the toad
immediately inflated itself again, so that it was standing on
the tips of its toes with its body taut and clear of the ground.
It evidently expected another attack by the snake, so I

moved it some distance to recover unmolested. – *J. A. Tudge*. [The description differs slightly from that of the usual defence posture, in which the body is tilted forward, with the forefeet and nose on the ground, so that the largest possible surface is presented to the attacker. – *B. C.*]

Cormorant retaliates

Cormorants feed in the sea not more than two hundred yards from my back door in Shetland and, when one of them catches anything difficult to swallow quickly, there is generally a great black-backed gull waiting to sweep down and snatch the prey. Two years ago I saw a cormorant struggling to swallow an unusually large fish when a gull attacked; the cormorant held on to the fish and a fierce tug of war ensued. The gull won but, because of its exhaustion after the duel and the weight of the fish, it could fly only slowly and with difficulty towards the shore a hundred yards away. To my amazement the cormorant took to wing in angry pursuit. Twice it overtook the gull in the air but failed to grab the fish. Then a second great black-backed gull suddenly swooped down and, though the first did its utmost to avoid the new attacker, the two came together and crashed into the water. At that moment the cormorant, still following, flopped down on top. While the gulls struck at each other, it recaptured the fish and with a final frantic effort swallowed it. – *John Peterson*

LITTLE OWL v THE REST by J. W. Fishwick

One evening in early summer I was watching a pair of song thrushes feed their youngster on the flat roof of the beautiful old hall where I worked. One parent, which I assumed to be the mother, then tried to chase it into flight over the parapet; but it evaded her and scuttled to the end of the roof, where it stood alone, protesting loudly. This attracted the attention of a little owl, which pounced from its perch on a chimney and landed on the young bird's back. Its beak tugged at the thrush's neck, and its round wings were spread as it dug in its talons. The victim's cries were pitiful, and the parents screamed their alarm. Both were soon attacking the owl, soaring into the air and diving in turn to deliver heavy blows at its head with their beaks.

While the battle raged, a family of blackbirds which shared the roof with the thrushes were led from the scene by their parents, also uttering panic cries. The owl showed no signs of releasing its grip, and soon the thrushes began to tire; but the adult blackbirds returned and took over the alternate sorties. This caused the owl to thrash the air with its wings, as their attack dislodged some speckled feathers from its neck. When they broke off, the refreshed thrushes moved in again and soon all four birds were climbing and stalling over the owl, diving whenever an opening appeared. The constant pecking and tugging brought the owl and its prey near the edge of the roof, where the captive began to put up a struggle too. This compelled the owl to get a fresh grip and rise slightly above its prey. The attackers redoubled their efforts and toppled it over the edge, but it did not release the thrush. After falling, it regained control and was soon flying away, climbing laboriously over the lawns towards the distant woods. The parents and their allies continued their attacks in the air and so interfered with the owl that it first flew parallel with the hall, losing height, then returned towards the roof, weaving between the trees in the park to confuse its pursuers. Obviously feeling the effect of its load, it made a rather flat turn near my window and headed for the woods, trying to gain height.

The thrushes and blackbirds were becoming weaker and more ragged in their assaults, and I did not expect to see the young bird again. But a crow, flying about twenty yards away among the trees just above roof-top height, saw the owl, immediately altered course and flew at it, turning it completely round in mid-air. The other birds held off as the crow gained height and closed from above. Landing on the owl's back, the new attacker used superior weight and wing power to fly it down to the forecourt and hold it there, pulling and tugging at its neck. I ran downstairs and, as I arrived, the crow flew up to join the other birds on the roof. I struck at the owl with a stick, knocking it partly off the young bird, which was still very much alive. I struck again, but the owl would not release its prey; it thrashed its way into the air over the lawns once more, its burden held in one talon only. I threw my stick, which missed the owl but went near enough to make it swerve and hit the top of a yew hedge. At last it dropped the thrush and escaped to the woods. I took the young bird, which had fallen to the ground, and gave it a drink by allowing a tap to drip on its beak. When I released it, the parents were soon feeding it again.

PHEASANTS IN A GARDEN by E. Faithfull

Ali and Solomon are wild cock pheasants, not quite aptly named because, while Solomon is glorious, Ali's glory is even greater. His breast shines a rich purple in the sun; his rump has a pale green sheen, spangled with gold; and his tail, an inch or two longer than normal, forks slightly, accentuating his oriental magnificence. Solomon owns the north end of the garden and the compost heap where kitchen scraps are flung. Ali reigns over the rest of the garden and the south and west sides of the house, where he comes to be fed on a bank in front of the kitchen window. An imaginary line from the south-east corner of the house, across the lawn to the shrubbery, divides their territories; and on most mornings there are parades and neck-stretching duels to preserve the boundary.

The battle started in earnest one spring, when three hens shamelessly flirted with each cock in turn, picking their way through the garden and into either territory with impunity. If one cock ventured too near the boundary line, the other would be there too. At first there would be a dignified pacing

back and forth parallel to the line, sometimes fast and sometimes slower, with pauses for pecking up ants and flies. If
one started to encroach, a duel would begin, each crouching
with neck outstretched along the ground, immobile except
for an occasional intimidating thrust. When each was within
his territory this was enough to establish rights; but if one
was over the line a fierce battle ensued. Thrusts became
sharper and quicker, culminating in standing jumps at each
other, implying the threat of spurs, until the intruder was
back on his side of the line. Whether the duel was fierce and
active or confined to neck-stretching, the birds gradually
resumed normal positions, after the loser had turned his
head away and pecked a few ants – an acknowledgement of
defeat that allowed the winner to do the same.

On one occasion I witnessed a grand drama. For three
days Solomon had had all the hens with him on the compost
heap, and Ali had stood about his kingdom looking forlorn.
As I watched, I saw him come on to the lawn and approach
the boundary line, where he paused; then suddenly he
marched quickly and determinedly towards the compost
heap. About twenty yards short of it he stopped, and a
second later he was streaking back down the lawn with
Solomon in hot pursuit. Ali was overtaken two yards short
of the imaginary boundary, and the fiercest of spurring
battles began, with jumping and stamping, until at last
he was driven back and had to acknowledge defeat. After
considerable dalliance the hens divided themselves, two
to Solomon and one to Ali. Thereafter the cocks were
even more alert, standing on tiptoe and crowing to assert
their rights, or racing to an encounter; and the battles became more frequent, fiercer and more prolonged. Solomon's
hens fed with him on the compost heap, but Ali brought his
to the kitchen window. He would feed first, making little
soft clucking noises of encouragement, then withdraw into
the laurels while she had her turn, finally escorting her
down the garden again. For a short while the hen appeared
only at regular times, morning and evening, and a little
later not at all. Ali continued to come regularly, but it was

a long time before we saw the hen leading three large and leggy chicks through some long grass. In the end all came to the bank by the kitchen window to be fed, the poults together and rather shy about coming forward from the laurels, the hen by herself and bold, walking to the edge of the bank and looking up to the window for recognition.

Stop go

The resident cock blackbird was foraging at the end of my garden, which is mostly grass broken up by rectangular flowerbeds. Presently a fully grown rat came up from an old rabbit hole and sniffed the air; then it began to work up the garden, casting here and there with short quick rushes. The blackbird stopped searching, ran a little way along a grass path and stood with head cocked. The rat followed, stopping short of the bird by a yard. The blackbird ran on again and stopped; the rat did the same. Up one path, down the next, neither cheated by cutting across the beds. So they continued until they reached the top of the garden when the blackbird, without fuss or hurry, flew up on to the fence. He gave no warning call as he would have done if frightened. The rat turned and went back the way it had come. – *Stella V. Mann*

I have been amused several times to see a cock blackbird and grey squirrel apparently playing together on the ground. The squirrel cavorted about and occasionally made short rushes towards the blackbird, which made no effort to fly but kept a safe distance; not once did he utter his alarm note. – *G. R. Oliver.* [This behaviour towards a potential enemy seems to be related to that used in territorial defence, when cock or hen blackbirds will manoeuvre at a distance, sometimes for minutes on end, without making contact. – *B. C.*]

Giddy-go-round

Sitting in my car one March day on the old drove road across Sheriffmuir, I saw through my field-glasses a movement in the grass and heather. A small rabbit was running in a tight circle about five feet across, hotly pursued by a hooded crow which, helped by its flapping wings, was moving jerkily but quite fast. After some five minutes the crow suddenly overbalanced and fell sideways. The rabbit ran a few more turns round the circle before it realized that it was no longer being chased, stopped and, with a flick of its hind legs, disappeared down a hole. The crow, disappointed of its prey, took wing into the haze over Allan Water. – *Stewart Campbell*

Not impressed

A young rabbit appeared in my garden and seemed to relish the clover growing in patches on the lawn. It called every few days, disregarding the sparrows, starlings, chaffinches and nuthatch which is a regular visitor. One day it was quietly eating clover when a woodpigeon landed about six feet away. The rabbit looked 'surprised-like' and rushed at the bird as if to annihilate it; but the pigeon was unmoved, even when the rabbit jumped clean over it. The rabbit then put on an acrobatic display, circling round the pigeon, jumping with all four feet off the ground and contorting itself in the air, making repeated feints. The pigeon watched the performance for about ten minutes before flying off. – *J. T. McLean*

Hickory dickory dock

During a late summer picnic in a Nottinghamshire wood we watched a robin take a bath in a puddle, then perch in an elder bush where it preened assiduously. Suddenly it began a series of attacks on something moving in the undergrowth.

This was a long-tailed field mouse which was climbing the tall hogweed stalks, presumably to feed on the seed heads. At each attack it scampered down the stalk, only to re-appear up another a few seconds later. Not until it had moved some way from the bush did the robin leave the mouse in peace to continue its diligent quartering of the hogweed. – *Elizabeth W. Douglas*

The best defence?

On a cloudy autumn evening our newly improved road-lamps did little more than emphasize the darkness pressing round them; but in a lighter stretch of road I saw a white cat crouching and jumping, occasionally putting out a paw at something. I hurried to prevent more torture, but there was no need: the cat was not the aggressor. A pygmy shrew had apparently been going about its lawful occasions, when the cat had interfered. Each time it poked out a cautious paw, the shrew would leap at its nose. The cat would then lift its head and retreat, followed by the shrew. A wave of my stick and a discouraging growl sent the cat through the hedge; but when I looked to shepherd the shrew carefully in the opposite direction it turned on me. I took a few steps backward, not wanting an inch-high assassin up my trouser leg. The shrew followed, and only the scrape of my stick on the road sent it after the cat. – *James Ker Cowan*

A pair of magpies which had nested in a hawthorn hedge were disturbed by a black-and-white cat sitting in the field nearby; and I watched for half an hour while they harassed it. One kept tempting it in front, as the other made constant attacks behind, severely tweaking its tail. The cat eventually retreated, waving its tail furiously. On another occasion, as I was taking a last look at my garden in the evening, I saw a little owl on top of a telegraph pole while my neighbour's cat sat on the wall below. Suddenly the owl dropped and knocked the cat fifteen feet down into my garden. – *Alan S. Bevan*

Buzzards in trouble

One day in summer I was walking through a Welsh valley when I heard a disturbance a little way up the hillside. A crow was flying back and forth across a piece of ground, cawing excitedly. Thinking some small creature must have been caught in a trap, I hurried to the rescue through brambles and thick undergrowth until I came to a small clearing. To my surprise a full-grown buzzard was flapping about with wings outstretched, apparently unable to rise from the ground. As the day was hot, I had neither coat nor gloves to protect my hands from its sharp beak and even sharper claws; and I was preparing to use my skirt when it raised a wing. I then saw that a second crow was gripping its right side just below the wing. The first crow continued to caw loudly as it flew overhead. By stamping, shouting and clapping I induced its mate to release the buzzard, which flew off down the valley like a rocket. – *D. McHattie*

In late spring I was lazily contemplating a buzzard which was circling over the Tamar Valley without apparent movement of the wings, now and again uttering its cat-like cry. Then, after a few tentative swoops, it dropped like a stone with closed wings. At the last moment it reopened them and, as they brushed the grass, its talons shot forward to seize a hidden object, with which it soared again; a little body was swinging from its feet. A few seconds later I realized that the bird was not rising with its customary ease: the wing-beats were losing their power and regularity, and they weakened until captor and captive began to fall. The descent was checked by a frantic outburst of energy, then the blurred mass was lost to sight beyond the far hedge of the meadow. Eventually I found the buzzard dead in the grass, and a weasel gripping its breast with meshed teeth; the bird's underparts were smeared with bright red blood. The farmer, who had joined me, confirmed that the animal would hang on until its teeth met. Although buzzards are recorded as taking weasels occasionally, they must run a great risk unless they succeed in dealing a death blow at their first stoop. – *E. Anderson*

Squirrel and stoat

One day in November my wife and I were out in woodland near our house in North Staffordshire when we heard cries of distress. A grey squirrel and a stoat, locked together, came tumbling down the wooded slope and a fierce struggle followed among the bracken at its foot. Try as it would, the squirrel could not free itself from the grip of the stoat. It sprang into the air again and again in attempts to shake off its attacker. After about fifteen minutes the squirrel lay still except for nervous twitches, and the stoat began to feed. Quite suddenly it stopped and left the body, which lay on its back, and I could then see that the heart was still beating very slowly. The stoat moved among the bracken, never more than six yards from its victim. After some little time spent sniffing around, it approached the squirrel again, seized the body by the neck and dragged it, a foot or so at a time, into cover. Monica Shorten, in her book on squirrels, mentions that they are occasionally taken by stoats, but it must be very unusual to see the combat. – *G. A. Lovenbury*

Terrier rabbit

Long after the swifts had departed I heard a sound like their screaming from the cliffs above Blue Anchor in Somerset. At first I could see nothing when I trained my binoculars on the cliff-top; then I caught a glimpse of two animals dashing about among the scrub. A second later a rabbit came hurtling down the steep slope, slithered, lost control and somersaulted several times on the rocks. I expected to find a battered corpse, but it recovered immediately and set off along the stones above the beach at a furious pace. Suddenly I saw in front of it a stoat, its black-tipped tail waving in the air as it fled. The rabbit jumped on to a larger boulder and paused, bolt upright. The stoat, a few feet ahead, paused too and, poking its head out from the shelter of a rock, faced the rabbit, which leapt forward. The stoat again fled and the chase was resumed, back up the cliff and among the hawthorns. The last I saw of the encounter was the rabbit patrolling on the cliff edge, like a terrier after a rat. – *D. Knowlton*

Not so funny

When our corgi bitch came into season for the first time in October, we took her by car into the country for a good walk. In a field we saw a number of hares, and the dog was soon after them among the turnips; but two began to chase her, and she came back to us, tail between her legs. We wondered whether it was her scent or hare-like appearance that stimulated them. – *Florence Hopper*. [Hares will run towards an unfamiliar sound or sight which is not immediately recognized as potentially dangerous. But dogs are regarded as enemies, and does with leverets have been known to attack and strike them with their forelegs. Cows may also be rapped on the nose when feeding too close, and there are records of humans being struck on the shin. As litters have been recorded in almost every month of the year, it is possible that there were young leverets among the turnips, and more likely that the corgi's attackers were does than bucks. – *Winwood Reade*]

No hiding place

In woods at Kimbolton, Huntingdonshire, my dog chased a grey squirrel which climbed the trunk of an old tree and plunged into a hole made by woodpeckers. There was a clamour of bumps and shrieks, and out came the squirrel headlong, pursued by a mob of angry wasps which followed it as it dashed away through the brambles. – *J. Dearnley*

Tug of war

A weasel dashed across a track where I was standing, and I saw a bird in its mouth. I ran after it, hoping to identify the prey, and it disappeared down a hole, leaving the bird, a skylark, just outside. I crouched by the corpse, and after a few seconds it was dragged to the hole by a wing; so I pulled it back by a leg, forcing the weasel to let go. The animal made another grab as soon as I had relinquished the bird, and once more we had a somewhat one-sided tug of war; this time I dropped it about six inches from the hole. The performance was repeated, and after each victory I pulled the bird farther away, until the weasel had to come right into the open, less than two feet from me. Finally, feeling it had worked hard for its dinner, I left it to eat in peace. – *Brian W. Burnett*

STOATS AT WORK by Raymond Hewson

One clear February day I watched a stoat hunting on the bleak Cabrach moorland of Banffshire. It was in winter coat, white except for the black-tipped tail and a wash of brown half an inch wide at the shoulders and nape. After each dozen or so bounds it sat upright with forepaws together, like a dog begging, and looked quickly all round with jerky head movements; or sometimes it paused with front legs extended and feet on the ground. By apparently random short rushes, it worked its way across three acres in twenty minutes. Grouse took to the wing when it came near them, but two mountain hares sat motionless in their scrapes in the peat. No prey was captured, and none pursued; I got the impression of rather speculative hunting by sight.

On the other hand, scent is clearly important. The first stoat I watched hunting in Scotland was following a mountain hare; both predator and quarry were conspicuously white in a Morayshire pine wood. The hare was moving slowly, the stoat following two or three seconds behind, casting about excitedly along the scent line. After four or five minutes the hare reappeared among the trees, ears erect, apparently stiff or tired. It ran along a broad woodland track and turned off into the trees. As it did so, the stoat came on to the track; and travelling in great bounds, faster than the hare, it finally picked up the scent and vanished among the trees. Here, as with a weasel I watched hunting a vole across an open field, there was apparently inefficient casting about after scent, compensated for by the predator's greater speed during pursuit and seemingly limitless energy. (In normal circumstances a mountain hare is much the faster of the two.) But sometimes a stoat is able to make more effective use of its nose. I followed the tracks of one parallel with, and always slightly downwind of, those of a rabbit along coastal sand-dunes and here there was no sign of casting about.

The stoat's method of killing a rabbit by biting the back of the neck is as widely known as the apparent indifference of nearby rabbits. I watched one bound five feet on to the back of a rabbit, which struggled, threw it off and moved, as if infinitely weary, about a yard. Four rabbits, the nearest ten feet away, sat and watched. When the stoat renewed its attack, the victim rolled over and lay still. The stoat got off and moved away. Ten seconds later the rabbit slowly travelled the six feet to a burrow; and the stoat followed,

jumping on and off its back and hanging on as it entered the hole. I walked up and saw the rabbit at a bend a yard from the entrance, with the stoat peering round it and sometimes coming a few inches towards me. Then the stoat retreated, until I could see only the green light reflected from its eyes, and began to tug its prey round the bend. This done, it reappeared, quite immaculate, licking its lips.

Stoats are very strong and able to carry half-grown rabbits with head held high. Pauses every ten yards or so may be as much to look around as to rest.

In much of Scotland and occasionally in northern England the winter coat, grown during an autumn moult, makes stoats much easier to watch. Although not all become white, even in north-east Scotland, like mountain hares they are often conspicuous against a winter habitat of heather moorland and peat. They become brown several weeks earlier than the hares in spring, after moulting later and probably more quickly in autumn.

As I was watching a stoat in winter-white hunt on bare ground beneath riverside alders in Banffshire, I noticed that a kestrel had flown into a tree just above it. The stoat moved off, occasionally disappearing among tangled herbage or briars. The kestrel followed, and again perched over it. It then occured to me that the bird was waiting to pounce on voles or mice disturbed by the stoat. After another move it did swoop to the ground, but I was unable to see whether it caught anything. At this point I lost the stoat

and was hoping that the kestrel would lead me back to it, to confirm my notion that the bird had learned to follow the animal to obtain prey. But it flew to the top of a dead alder and stayed there.

Before the spread of myxomatosis rabbits were the stoat's main food, and I did not see one with other prey. Since then I have twice watched stoats kill or carry brown rats – powerful and heavier animals, which will fight back when cornered. Once I disturbed a stoat lying alongside a rat as if grappling with it. The stoat ran back to the rat-infested corn stacks by the roadside and, as it had not reappeared five minutes later, I went over to examine the rat. There was only a small amount of blood at the nape and a little fur missing, but the animal was moribund and quite still.

Recently in Wester Ross I was walking along a snow-covered ridge at about two thousand six hundred feet. Steep and craggy hills, sharp in the winter sunshine against a blue sky, stretched into the far distance. A mountain hare had skirted one of the great corries and, venturing out along a snow cornice, had broken through and rolled down the hillside until it was able to secure a footing and come up another way. A golden eagle flapped slowly over Loch Duich, a raven croaked from the summit cairn, and there were marks of fox and ptarmigan. As I moved towards the highest point of the ridge, clear in the firm snow were the tracks of a stoat. In this harsh and wintry place voles were its likeliest prey, and perhaps a ptarmigan or mountain hare. I realized quite suddenly how much, over the years, I had enjoyed watching this bold and graceful animal.

Tooth versus claw

As pendants to Raymond Hewson's article follow five notes which suggest that competition between predatory birds and mammals may be of quite regular occurrence. Sometimes the bird wins, sometimes the mammal – though it may be deprived by a bigger one, as in P. Hale's complex observation; or man may appear *deus cum machina* like Desmond Hawkins, the 'father' of the BBC's post-war natural history programmes, on his Dorset road. – *B. C.*

Driving through Dorset in October, I turned a bend and

unwittingly added a third danger to a mouse in the middle of the road. It was already threatened by a stoat running out from the grass verge and by a kestrel diving from above. The kestrel and I braked hard, the bird checking with a flurry of wing-beats a couple of feet off the ground and just short of my offside wing. As I stopped, the stoat snatched the mouse in front of the car and dragged it to the verge. The kestrel swung away; and I drove on, startled by the coincidence that had brought the four of us so dramatically together. – *Desmond Hawkins*

A patch of white and daffodil-yellow against bleached grass tussocks caught my eye, a pair of wildly beating wings showed for an instant, and my binoculars revealed a hen kestrel gripping with yellow talons the white underfur of a young rat. The bird was dragging this awkwardly across an open field within sight of a farm, pausing repeatedly to scratch her face with a claw. Then I saw a stoat circling the spot where the rat had lain a moment before; it picked up the trail and reclaimed its prey. The kestrel took off and hovered low down before banking steeply into the nearest tree, where she continued to preen and scratch, evidently in reaction to her anxiety. By then the stoat had carried the rat some twenty yards to cover by the river's bank. I continued to watch, and down the hill from the farm came a second stoat holding a rat by the head in its jaw. The kestrel stooped from her tree, diving and circling three times, but the stoat hardly swerved except to negotiate clumps of weeds, until a thick patch checked the rat's body. The bird struck again but rose baffled as the stoat vanished with rat underground. – *J. E. Hemingway*

On one of those rare warm days in March, when the sky was a clear turquoise and the air full of the song of larks, I was walking with two beagle puppies across the big water-meadow to the river, when a flock of rooks alighted on the opposite bank and, chattering excitedly, formed themselves into a circle. I slipped the puppies on the lead and, as I moved cautiously forward, a sharp squeal rent the air. The rooks were gathered round a stoat and rabbit. The beagles quivered with excitement when the rabbit, kicking violently, squealed again. Suddenly the rooks whirled cawing into the air and settled on nearby trees; and the stoat, abandoning its twitching victim, disappeared swiftly into the undergrowth. One of the beagles began to growl. Then I saw him – a magnificent dog fox with a large white tag on his fine brush, his russet coat gleaming in the sun. The puppies yelped and pulled on the lead. Quite unperturbed the fox picked up the rabbit and stood looking across the water at us before walking majestically and unhurriedly away. – *P. Hale*

A wood mouse ran across a metalled road in the Forest of Dean and disappeared into the grass close to my parked car; hard on its heels was a weasel. After doubling back, the mouse moved more slowly and uncertainly alongside the verge. Suddenly, from its concealed perch in a lofty oak, a jay swooped down and dealt the mouse four sharp blows with its bill, picked up the victim and returned to its hide. The whole episode was over in about half a minute. I have seen magpies, working as a pair, deal with a toad and a young stoat like this, and wondered whether the sight of the hunting weasel had alerted the jay to the chance of an easy meal. – *R. J. Jennings*

There was an almost imperceptible movement of grasses as the field vole darted away on the primrose-dotted railway embankment. A few yards off a weasel moved sinuously in the same direction. Overhead a kestrel hovered, watching every movement below. A permanent-way man, walking along the single-line track, drew steadily nearer; and the sound as he knocked the loose wooden keys with his hammer became increasingly audible, until a particularly loud clang caused the weasel to raise its head and turn from its objective toward the threatened danger. At that precise moment the kestrel, quick to spot the distraction, stooped on the vole, seized it and bore it aloft. – *S. C. Marshall*

Part Five

Routines and Displays

Introduction

The study of animal behaviour may be divided into that of routines, concerned with such daily activities as food-searching and eating, washing and cleaning, contacts with other animals, resting and sleeping, and the formalized movements known as displays and particularly associated with birds, because theirs are the most spectacular and easily studied.

But where does 'play' fit in? Is it routine or display, and what is its purpose? T. J. Richards, whose article opens this Part, describes actions by ravens and crows which seem purely playful, but he suggests that the peregrine's mock attacks on other birds serve to keep it in trim for hunting. Greta Phillips' crows at their follow-my-leader were enjoying a form of play observed in other species and superficially very like the repetitive games of human children.

Anting, sunbathing and smoke-bathing are responsible for some grotesque attitudes by the birds indulging in these mysterious activities, which are now considered to be concerned with care of the feathers, a reasonable suggestion in view of the vital importance of its plumage to the well-being of every bird. Anting appears to be confined to birds in the passerine order, which includes all our familiar song-birds. Dr Hans Löhrl, the distinguished German ornithologist who directs the bird observatory at Radolfzell, provides a photograph showing the typical anting posture, assumed by a tame jay which could not consummate the behaviour. C. H. Cooke reviews sunbathing, which occurs throughout the class Aves.

P. G. Brade's car-buzzing pigeon and George Hamilton's reported method of ridding a trawler of herring gulls attract comment by Derek Goodwin, author of the recent monograph on the pigeons and doves of the world, and Professor Niko Tinbergen, whose comparative studies on the behaviour of different kinds of gull have made an outstanding contribution to the analysis of animal behaviour.

Elizabeth Middleton's roosting starlings may be compared with Susan Cowdy's gulls in Part Eleven; John Platt's wire-hanging kestrel suggests a high degree of intelligence: perhaps readers who can lift their eyes occasionally from the road will look for other instances.

I have brought together a number of observations on image-fighting that have been published over the past few years. As a boy I was completely baffled by seeing a pair of redstarts attacking a back window of our Highland cottage. Now the mirror is a regular piece of equipment used to promote aggressive displays and attacks by larger birds, as described in *Birds Fighting* by Stuart Smith and Eric Hosking. It is possible that the hub caps of cars, which are often the targets of these attacks, exert by their distorted image what ethologists call a supra-normal effect on the bird and prolong its stimulation. In nature the territorial invader normally retreats when threatened.

Janet Bodman's delightful goldcrests and Henry Tegner's engrossed woodcock lead naturally to the subject of the next Part.

BIRDS AT PLAY by T. J. Richards

One stormy November afternoon I stood on a cliff-top near Branscombe in Devon. A half-gale was blowing and a solid rush of air streamed up the face of the cliff, tearing at grass and stunted thorn bushes. Just above the shingle beach two pairs of ravens were sporting. First they would sail with half-closed wings and dangling legs full into the teeth of the wind, each pair keeping close together as they undulated

slowly forward, like sea birds riding the waves. Then, after they had progressed some three hundred yards, the four birds would turn and streak downwind at top speed till, on reaching their starting point, they wheeled back into the wind and repeated the performance. They reminded me of children slowly climbing a hill with a toboggan to enjoy the swift exhilarating descent. I watched them at this game until I could no longer stand the buffeting of the wind.

Carrion crows have a well-marked type of play which they conduct above the windswept brow of a hill. At four o'clock one October afternoon I came on a party of sixty or seventy sporting above a ridge which overlooks the Otter Vale. Most of the birds were hanging in the wind only a few feet above the bracken, sometimes diving, twirling and making playful sallies at one another, while uttering the gurgling snarls crows use when mobbing buzzards. Others were resting on the topmost branches of small birch and pine trees. Now and again one of the birds in the air hovered clumsily with its hanging legs a few inches above a resting bird, as if about to settle on its back.

This would result in a brief squabble, the bird in the tree defending its position with lunging bill, until one or both joined the milling crowd riding the wind. The birds were constantly changing position, rising and falling like huge wind-blown leaves. Once or twice a crow dropped to the ground and rose with some object – a small stick or a piece of bracken – in its bill. Then, while poised in the air at a height of twenty or thirty feet, it would repeatedly transfer the object from bill to foot and back again before dropping it. One bird tried to rise with a large stick in its bill, but the burden was too great and was released a few feet from the ground. The fun continued for nearly half an hour; then the birds dispersed, some flying away over the valley, others along the hillside. One passed overhead still carrying in its bill an object which I could not identify at the distance.

About a fortnight later, at the same spot and at much the same time, I again found the game in progress. Only a dozen or so crows were taking part, but the pattern of their play was preciscly the same as before: the gliding, swooping and diving; the 'spectators' at rest on the tops of small trees, while others appeared to covet their perches; and the juggling with objects picked up from the ground. After a few minutes the party was rudely broken up by a pair of ravens which came flying low along the hill and straight through the flock. The crows at once scattered with croaks of protest, one of them making a half-hearted stoop at the spoilsports. After wheeling once or twice among the indignant crows, the ravens continued on their way, conversing in deep *korks*.

The most interesting part of this corvine play is the passing of inedible objects from bill to feet. It is unusual for birds other than hawks and owls to carry objects in their feet. Yet the practice may be more common in crows than we think. I once saw a crow fly from a farm plot with a large potato spiked on its bill. Before long the bird seemed to find its heavy burden unmanageable. Lowering its head, it transferred the potato to its feet and carried it thus, with dangling legs, for about one hundred and fifty yards before pitching in a meadow. Nor is juggling with objects in mid-air a form of amusement confined to crows. I have seen a rook, followed by four of its companions, circling with a dead leaf which it repeatedly passed from bill to feet.

The peregrine's sky-play is perhaps of a grimmer nature. For although the bird is evidently revelling in its mastery of the air, its diversions always appear to be directed towards perfecting the art of killing. A raven or buzzard often becomes an unwilling participant in the game, being subjected to persistent mock attacks while running the gauntlet with croaks or wails of protest. If no other sport offers, a pair of peregrines will play together, each using the other as a practice target. As one bird streaks across the arc of the sky, hurtling like a bomb towards its mate, it seems that nothing can prevent the impact. But at the last minute the falling bird throws up and soars again to the zenith, while the other flicks lightly aside.

On an autumn morning I watched a peregrine stooping time after time for no apparent reason other than pleasure. After each stoop the bird would soar to about a hundred feet and then, on set curved wings, ascend on an air current as though drawn up by a string. At its pitch, five or six hundred feet above the cliff, it looked like a large swift fixed motionless against the drifting grey cloud. With binoculars I watched it turn and dive obliquely, giving a few sharp flicks of the wings to gain impetus. Then, with tightly closed wings, it dropped like a pear-shaped bolt, looming larger and larger in the field of the glasses until it passed with a loud swish thirty feet above my head. The descent took roughly four seconds.

Birds being what they are, it is natural that most of their play should be of an aerial nature; but I once watched an immature herring gull amuse itself on the sand of an estuary, The game consisted in pursuing a dead leaf blown across the sand by the wind. On catching the leaf, the gull would hold it by the stalk for a moment or two, allowing it to flutter in the breeze. Then it would release the leaf and chase it as before. The game ended when the leaf blew into a pool. The gull waded in, retrieved its plaything and dropped it several times into the water, as if expecting it to run away again. But the leaf had become inanimate, and the gull, losing interest, began to preen.

Image-fighting

Seldom is a bird attacking its reflection caught in the act so neatly as the female pied wagtail photographed by H. J. Vosper at his camp site by a sea loch in Argyll (Illus. 12–14). She had been seen about, but not until the beginning of the second week did he find droppings early one morning on the nearside front wing of his car. Later, sounds of scraping and fluttering drew attention to the wagtail posturing before the wing-mirror. 'Crouching down, then stretching up to full height, occasionally being blown off the slippery surface, she continued the act for several minutes until, infuriated by the mimic in the mirror, she flung herself forward and pecked vigorously at it. Repulsed, she flew over the rival's head to land on top of the mirror's rim and gaze around.' This performance was repeated again and again, despite discouragement, so the mirror was masked; but in succession the bird attacked the offside mirror and the backs of both. When they were hidden, she fed on the ground for a short time before discovering a hub cap. Eventually all four caps had to be covered; but the following day, the party's last in camp, the wagtail renewed her attack by assailing the back bumper-bar.

Another persistent attacker of hub caps over six weekends was a cock blackbird. When disturbed, he either retired under Molly Sole's car or perched nearby and scolded loudly. One morning, finding the cap smeared with sticky red, she feared the worst, but soon surprised the bird returning with a morello cherry in his beak; evidently he was trying to feed his reflection. Dr D. W. Snow thinks he may have been feeding a brood elsewhere and was simply confused, so that he was trying to feed the wrong object.

Two crows attacking a french window of Betty Cohen's house at Sully near Cardiff rapped sharply on the glass several times, cawing loudly, before flying up to scratch it with their claws. On landing, they set about each other so fiercely for half an hour or more that feathers flew and the window became smeared with dirt and drops of blood; some also fell on the concrete path. Fighting continued every morning for a week, then suddenly the pattern changed. The crows would perch on the low, almost bare branch of an ash about twenty yards diagonally from the window. After cawing and bowing, they launched themselves at this, striking it one above the other with a good bang, then dropping to the ground to fight as before. The placing of various articles in the window did not deter them but, after a week, a sheet of polythene tacked over it stopped the attacks. Derek Goodwin comments that it is not uncommon for even paired birds to attack each other if they are prevented from venting their anger on its cause – a form of redirected aggression.

Another incident, involving both crows and magpies, was reported from Surrey by Brian W. Burnett. 'Disturbed by the noise, I looked out of my window one day in May to see two crows on an empty chalet about a hundred yards away. One was on the roof, and the other flew up to a window and banged against it. For the next hour or so I watched them, sometimes at different windows, sometimes together, cawing incessantly as they banged against the glass. During their occasional short breaks they strutted about on the ground or roof. A pair of magpies flew down behind them and were ignored, even when one began to pull at the tail of its big relative; but after a minute the crow chased it away. For about an hour the magpies made frequent playful attacks on the crows, which were still very noisy, though their tormentors remained more or less silent. After the magpies had flown off, the crows continued to bang at the windows, so I chased them away to see what would happen. They soon returned and were still at it when I had to leave, four or five hours after I had first heard them.'

Other birds reported image-fighting in recent years are rooks, chaffinches, grey wagtails and a house sparrow watched by Mollie Ingram in Somerset. 'In April 1963 I first noticed a male house sparrow banging against a bedroom window of my neighbour's house. It kept up its attacks from early morning until late evening except when, at my suggestion, the curtains were drawn. At times the bird changed to another window but, with a break of five or six weeks in the autumn, it continued its visits until the following June; then it disappeared, coming back at the end of March this year to the same window. It kept up its attacks for a month before vanishing again.'

The habit is not confined to European birds, and several readers in Africa have sent us records. In East London W. H. Archer suffered assaults for several days on the bonnet and windscreen of his car by an African wagtail. A bokmakierie shrike also attacked his workshop window for weeks, until it so fouled the glass that no reflection could be seen; sometimes it dive-bombed from its perch on a telegraph pole. The arrival of a mate ended its performance. From Kenya, Mary Rickman writes of attacks on windows by a male golden weaver and, less violent, by male and female rufous sparrows. The habit evidently proved fatal to a robin chat which she found dead under a window; it had previously attacked a car windscreen. In Rhodesia a Kurrichane thrush would keep up its 'tinker's tap-tap-tap' for hours on the hub caps of Chris Mears' old car. When she moved house and bought a new car with higher, more convex hub caps, the local thrush 'roused itself into an absolute frenzy, leaping up with feathers flying to reach its reflection'. Both sexes of scarlet-chested and double-collared sun birds have shown interest in her windows, usually 'chortling and preening' from the nearby golden-shower creeper, but sometimes hovering against the glass, 'taking odd pecks at their reflections'. A neighbour was woken early by a male puffback shrike tapping at the window and turning himself into 'a fluffy white snowball'. – *B. C.*

Ants but no anting

The tame jay (Illus. 15) assumed full anting posture when I put ants on the ground near it. Sometimes it almost fell over as it tried to pick up those crawling over its feet. But I never saw it complete the routine by rubbing them on its plumage. – *Hans Löhrl*

Follow my leader

The three fledged young of a pair of carrion crows, which nested on the outskirts of London, settled on the ridge of a roof opposite our house while the parents flew off to forage. One juvenile carried what seemed to be a crust of bread, which it allowed to fall and slide down the roof to the gutter. The bird eyed it, then proceeded to toboggan down the slope after it, but fluttered up to the ridge again without the crust. After contemplating the roof briefly, the crow slid down again, soon followed by the others. They kept up this kind of follow-my-leader game, down the slope and fluttering back to the ridge, for some minutes before settling to rest and await their parents' return. – *Greta D. Phillips.* [A tame raven repeatedly rolled down a snowy bank; there are also records of starlings sliding down snowy roofs, and of eiders shooting the rapids of a river in company. – *B. C.*]

BIRDS SUNBATHING by C. H. Cooke

One June morning more than a hundred house martins were gathered on the sloping roof of a house overlooking my garden at Hitchin. The sun was shining directly overhead from a cloudless sky, and they were reclining with wings outstretched and bodies flattened against the roof, taking full advantage of the powerful rays. Single birds flew off occasionally but, after fluttering round for a minute or so, returned to continue sunbathing. Through binoculars I could see their feathers puffed out and their bodies slowly palpitating; a few had only one wing extended. This lasted for almost an hour, until the sun moved round and left part of the roof in shadow. Although the warm spell continued for several days, I did not see the martins sunbathing again.

Since then I have been on the lookout for birds sunbathing. Two years ago in July I saw a great tit on a concrete path in my garden with both wings fully extended and body pressed close to the hot path. Last year I watched a lapwing lying on the bank of a disused gravel pit, its body pressed against the hot mud with both wings outstretched; and it remained in this passive position until disturbed by a sudden noise. When the winter sun shone brightly some house sparrows were sunning themselves with spread wings on a roof. The other birds I have seen sunbathing include bull-finches, starlings (Illus. 16), song thrushes, chaffinches and robins; and there are records also of red-backed shrikes, swallows, wrens, nightjars, barn and tawny owls and long-tailed tits.

Song birds have a characteristic sunbathing posture; they lean sideways with feathers erected and the wing and half of the tail facing the sun spread out. The beak is usually open, and the eye towards the sun wholly or partly closed. If the sun is overhead, both wings may be fully extended and the tail fully spread. Pigeons behave similarly, but sometimes lift one wing to allow the rays to reach the side of the body. Game birds, gulls and some waterfowl have been seen with wings held a little way from the body and spread, and also sitting with backs to the sun and wings held loosely on either side of the body to expose the back and rump.

PIGEON PATROL by P. G. Brade

About noon on a Saturday late in January I was driving down the A34 towards Trentham, about two hundred yards south of the traffic lights at Hanford. There was a fair amount of traffic, and south-bound vehicles were travelling at just over forty mph, regularly spaced some thirty feet apart. Directly in front of me was a light van. Suddenly a slate-blue pigeon appeared from above my car, swooped down and flew parallel with, and about five feet from, the nearside of the van. My first thought was that the bird had been struck by a car and was not in full possession of its faculties, but it was soon apparent that its actions were deliberate. Over the next quarter of a mile it 'dive-bombed' the van. Once, when it swooped over the bonnet and out of view, collision seemed inevitable; but a split second later it reappeared on the nearside. It transferred to my car, keeping station until I turned into a café car park near Trentham Gardens; then it continued southwards at un-undiminished speed about ten feet above the traffic. Seconds later I saw it among the north-bound traffic. It must then have turned left into Whitmore Road, where my parents saw it soon afterwards pursuing a vehicle towards Hanchurch. [Tentative explanations of the bird's behaviour are that it had been fed habitually from a vehicle, or even hand-reared by someone who took it about by car. A pigeon 'paired' to a delivery man used to go the rounds with him in his lorry. Pigeons normally descend in a way that suggests dive-bombing only in a place they know well or to avoid a bird of prey. So this one was probably making an attempt to alight, which suggests some familiarity with vehicles. – *Derek Goodwin*]

GRANDFATHER'S WAY by George H. Hamilton

When staying at Tenby in Pembrokeshire I spent a day at sea with a small trawler. We were catching fifteen to twenty stones of fish with each trawl, and the business of sorting and gutting began at once after emptying. The deck was covered with fish and soon with blood and entrails, which attracted a swarm of herring gulls. To put it mildly, they were a nuisance; so a member of the crew turned quickly from his work, caught one and proceeded to cover it with blood and entrails. He then released the bird which flew off, followed by all the others, to alight on the sea about fifty yards away. As it cleaned itself and preened its feathers, the others swam round in a circle and did not, as I had expected, attack it. Then all departed, and they did not return that day; nor did any other gull appear. The fisherman told me that this was the best way to get rid of gulls; his father and grandfather had used it, though he did not know why it succeeded. [Dr Niko Tinbergen, author of *The Herring Gull's World* in the New Naturalist series, writes: 'Although this particular observation is new to me, it fits in with others we have made. If a gull moves, calls or looks as though it is being seriously threatened, or if it is seized or wounded by a predator, this response occurs: first attraction to the victim, then scattering or moving away from the place of the incident. It also appeared when we faked the effects of an attack by putting out stuffed gulls in odd postures or by playing back recorded distress calls; and we have seen it when one of two fighting gulls is wounded or moves awkwardly. The more we learn of the reactions of gulls to predators, the more consistent the general picture that emerges: the stimuli are either those provided by the predator itself, or which emanate from the victim, or they may be the alarm calls and behaviour of other gulls in the vicinity. If the predator is present, the birds mob it; if it is not, as they discover on approaching, they leave the place where they saw or heard the victim. In experiments on the

effect of distress calls or other aspects of a bird in trouble
we find that repeated administration of the stimulus leads
to a gradual waning of the response. This is why the
application of devices such as scarers, for example on air-
fields, has only limited success. My guess is that, if the
fisherman's ruse were applied too often, the gulls would
gradually get used to it; but it would be worth finding this
out.']

MURMURATION OF STARLINGS
by Elizabeth Middleton

On the seventh of November I took some friends for a drive
over the Coniston fells to Hawkshead. As we were returning
down the west side of Esthwaite Water one of the party
exclaimed, 'What an odd time to plough a field!' We looked
across to what appeared to be a stretch of black earth, and
almost at once, like Birnam Wood, it began to move. The
whole field lifted in waves, and a cloud of starlings passed
over the car, like locusts, to a pasture on our right. The
noise was terrific. Sheep which had been peacefully nibbling
now bleated plaintively in protest and scurried up the fellside away from the black invasion, while the horde of
starlings pecked continuously with incessant chattering.
When they rose again, to swoop low over our heads, their
droppings rattled on the roof of the car like hailstones.
They swept downwards, and the green sward became a black
carpet stretched in a huge wedge across the field. There was
more pecking for a few minutes, until the top of the wedge
peeled off to join the mass at the other end. Then suddenly,
in a vast uprising, the birds made for a tall beech in front of
us. The elegant and shining-fingered tree immediately became a black blob, as the swaying mass crowded there for a
moment before plunging headlong to the field, where it
settled in a long spearhead. We were wondering what
the next move would be when, in a twinkling, the tip of
the spear sped away, followed by the rest of the shaft.

Filling the air with noise and darkness, the starlings vanished in their thousands across the lake. It was a wonderful sight; but we had to spend two hours cleaning the car.

Labour-saving

A kestrel was hovering over the grass verge of a road through the Mendips. It hung motionless, head to wind, with the claws of one foot negligently hooked round a telegraph wire, about midway between two poles; the other foot was half retracted. Thus supported by the breeze and anchored by the wire, which was raised several inches above its normal position, the bird had no need of frequent wingbeats to remain in place. – *John E. Platt*

MINUET by Janet Bodman

At the end of April a miniature cascade of notes announced the return of the goldcrests to their usual high position in our old yew tree. The voices moved closer, and I caught the flick of wings in the adjacent prunus: the cock was displaying to the hen, hidden in the edge of the yew. When he turned my way I could see the brilliant upstanding orange crest. His whole body was swollen with excitement as he prinked and pranced with swift staccato movements from one twig to another, up and down a small stretch as though on a diminutive stage; his turns and flutterings were all in silence, with occasionally a few notes from the hen. After about ten minutes of this concentrated ritual, the hen came into view; she looked greener above and darker below, her crest smaller and yellower. She hopped into the prunus a little below her partner and began a series of angled movements at the same pace. He responded, and together they danced a fraction nearer to each other, wings half-fluttering, tails spread, their actions swift and delicate. Then the spell broke and the cock flew away.

ROUGH COURTSHIP by Henry Tegner

About one o'clock on a cold May afternoon I entered a little moss-covered glade near Loch an Eilean in Inverness-shire. Between the trunks of the birches I saw two dark brown lumps leap into the air, and through binoculars I was able to identify them as woodcock. They faced each other, squatting on the ground, then suddenly sprang into the air like acrobatic dancers, seeming to lean backwards as they did so, and extended their feet towards each other, as I have seen gamecocks do in Spain. Sometimes the pair settled on the mossy ground, circling until one tried to spring on top of the other; or one bird would strut with trailing wings like a blackcock at the lek. I was able to approach to within ten yards and watch for at least five minutes. I am convinced they did not see me, because one, while circling the other, came within a few feet of me. In the end one took off, flying low towards the pines by the loch; its companion followed almost immediately, and I last saw them chasing

each other among the scattered trees. They were almost silent, except for a light guttural noise when one pitched close to me. In spite of the appearance of combat, I believe this was the beginning of the nuptial display, which has strong elements of aggression.

Part Six

The Crisis of the Year

Introduction

The crisis of the year is, of course, the breeding season, for success then is vital to the survival of every animal. The notes selected start with one of the humblest examples of parental care – among shieldbugs – and progress by way of a social insect, A. C. Hilton's cedarwood thieves, to the devotion of many birds and mammals when their young are threatened. Dr Ernest Neal, our leading authority on badgers, writes of an unusual record of breeding above ground and also supplies a classic photograph of multiple nest-building by a song thrush.

Frank Reeman describes a unique experience with a mole, and short notes on the depredations by jackdaws, house sparrow and starling in search of nesting material recall the opportunists of Part Three: such behaviour is consistent with the successful careers of these almost ubiquitous birds.

The various rescues of their young by wood mouse, moorhen (with apparent anticipation of danger), weasel and swan (with some human help) lead to illustrated notes by Bobby Tulloch and Phillip Glasier portraying the fortitude of birds that nest in the open. Stone curlews survive on the chalklands of the south, though somewhat precariously, but R. P. Gait's red-backed shrikes have vanished from their haunts near Bristol, where he photographed them.

The part ends with three notes about the odd man out of our breeding birds, our only brood-parasite, the cuckoo.

A MOTHER'S SHIELD by Ray Palmer

The social ants, bees and wasps and the earwig are not the only insects to show parental care. There are others in an unexpected group, the *Hemiptera* or bugs. Charles de Geer noted in 1764 that the female birch-leaf shieldbug (*Elasmucha grisea*) guarded her eggs and young, but his observations were not confirmed for nearly a century and are still not well known. So I was pleased to find one of these shieldbugs on a birch leaf in bright sunlight, brooding a large batch of eggs which appeared to be laid in diamond formation, so that her body would almost cover them (Illus. 9). She would not leave them, only settling down more closely when I touched her. The mother remains over her eggs for at least a fortnight and, when the young are hatched, continues to guard them for the first ten days. Her hard body must be a good protection against parasites and predators, though it has been suggested that the cannibal instincts of the male are the chief danger.

Later she moves about over the foliage, and the tiny bugs run after her like chicks following a hen, clustering beneath her when she stops. They keep contact with her and each other by means of their antennae. If one is attacked by an intruder, she will come to its rescue, touching it with her antennae, walking round it and sometimes placing a protective leg over it. Until recently the habit of guarding eggs and young was believed to be peculiar to the birch-leaf shieldbug; but we now know that the rare bilberry shieldbug (*E. ferrugata*), the deadnettle shieldbug (*Sehirus bicolor*), the forget-me-not shieldbug (*S. lactuosus*) and certain American species behave similarly. Since little is known about the life-histories of this group, further research may show that maternal solicitude occurs much more widely.

STOLEN SILVER by A. C. Hilton

When several years ago I erected my cedarwood garage, it soon took on the silvery hue of untreated wood that has weathered. The summer before last, on returning from a fortnight's absence, I noticed that several square feet of roof had lost their silvery appearance, and closer inspection revealed that the sides of the building had also suffered. I discovered the cause while working in the garden: worker wasps were busy skimming the surface of the cedarwood with their mandibles. When half a dozen were at work I could distinctly hear their rasping. It was surprising to see how much they managed to remove at a sitting, before disappearing over the wall to enlarge their papery nest. Early last spring I observed a queen wasp similarly engaged, evidently intent on constructing the first few cells in which to lay her eggs; and throughout the summer wasps have been at the garage, which is now variegated, dull patches intermingled with silver. By mid-August the wasps were not only rasping away at the garage but were gathering up the carcasses of the drones driven off by the worker bees. The hives were strong and the wasps did not attempt to enter them; they seemed content with a meat diet, perhaps an indication that their colonies still had many grubs in them.

Trial run

Early in August two half-grown red squirrels started to build a drey about eight feet up in a clump of philadelphus in an Aberdeenshire garden. Travelling together on the ground, they made five journeys in one half-hour to a copper beech about thirty yards away, returning with mouths full of dead twigs from under the tree. Coming and going they kept up the continuous chatter and chirruping which had first caught my attention, and they completely ignored me. During a sunny spell at the end of September I saw them gathering fluff from the seeding willowherb and carrying it to their drey. – *Kay Shimmer*. [Red squirrels born early in spring might well be mated by August; the low height of the drey suggests a practice nest. – *Monica Shorten*]

MIDWIFE TO A MOLE by Frank G. Reeman

We were wandering through a strip of woodland one after-noon in May when we noticed a rustling among dead leaves. I cleared a circle round the spot with a stick and poked about until a mole made a break for freedom, and I was able to catch her. We installed her in a rabbit-hutch, filled the nest compartment with a box of firm soil and covered the floor of the other half also with a layer, joining the two with a small wire-netting tunnel; and we added a little hay for bedding. To our surprise the mole proved docile and adaptable; she soon learned to come scuttling along the 'run' when we rattled the food tin, and she took worms from our fingers. If stroked, she stopped eating only long enough to remove the offending finger with an upward thrust of one forefoot. She was rather less than three and a quarter ounces and ate more than her own weight of worms on the first day. That posed a problem, because there had been no rain for weeks and the soil was dry; but she would accept lean raw beef as a substitute, having first finished off the worms we provided.

Three or four days after her capture we opened the soil box to find two tiny pink moles on the surface, one with its head bitten off; the other we removed and kept warm. When we put the mother on the grass, she sniffed around, accepted a worm and then started to chase her tail, uttering low-pitched yelps. We saw she was giving birth to a third baby, assisting delivery with her muzzle. When she had cleaned the pink and wrinkled offspring she nudged it, but it did not respond. She then cleaned herself, nosed the baby again and, when it failed to move, bit off its head and turned to burrow in the earth.

We fed the remaining young one (Illus. 17) at intervals with glucose on a fine brush; it could burrow effectively between the fingers of a tightly clenched hand but died after eleven hours. Would the mother have reared it, if we had left it with her? We were afraid that, in captivity, she would mutilate it with her sharp teeth, as she had the others. She stayed with us for two months after the birth, lively and in good health, until one night the cage door was not fastened and she slipped away.

BADGER NESTS ABOVE GROUND
by Ernest Neal

The badger has been described as the oldest landowner in Britain. It roamed the deciduous forests of southern England long before they became isolated from the Continent. Bones estimated to be sixty thousand years old were identified in a Mendip cave; and badgers are living there today. Some sets mentioned in the Domesday Book are still occupied. So the badger is certainly conservative in its housing arrangements.

The breeding nest is generally made underground in a hollowed-out chamber not far from an entrance to the set, and in autumn the parents fill it with bundle after bundle of hay and bracken. The time of the cubs' birth varies; in the south of England the peak period is the first half of February, though litters in January and March are not uncommon. When the cubs (usually two or three) are very small the mother curls up round them to keep them warm inside the heap. Heat from her body soon raises the temperature, which is maintained by the bedding on the hay-box principle when she leaves the set to feed. A maximum–minimum thermometer put down a set in February on the end of a long springy wire recorded $-10°$ C ($14°$ F) more than outside; and it is unlikely the actual breeding chamber was reached, so the temperature in the nest may well have been much higher.

What goes on in the normal underground nest is hard to discover, but in March last year I was told of a sow which had bred above ground – a very rare occurrence, and particularly interesting because she had chosen a disused outhouse on a farm near North Curry in Somerset. When I arrived in the late afternoon the farmer, Jack Richards, showed me a large lean-to shed butting on to his barn. It had been used for storing rough timber planking, and not until some of this had been moved about a week earlier had the nest been discovered; each time the sow had been watched she had taken no notice. We opened the door and stood quietly just inside the shed; it was rather dark because there were no windows, but we had a torch. As we listened, I was astonished to hear the unmistakable whickering of small cubs coming from the far side. I thought we made enough noise climbing over the loosely piled timber to scare any badger away, and the wickering stopped, so I switched on the torch and peered down between the planks. The sow badger was lying at full length on top of a great heap of hay. The light made her eyes sparkle, but she did not move; then she turned slightly on to her side, and I saw she was suckling a cub.

Hoping for a photograph, we moved some of the planks carefully to give an uninterrupted view; but the cub was no longer to be seen, because the sow had turned over again. After some time she slowly stood up, shook herself and without undue hurry or concern made her way between the planks into the open shed next door, though I did not see her go into the field. The cub had buried itself in the heap of hay, but I moved the top covering gently with a stick: in the centre were two small cubs curled up nose to tail. Their eyes were open and the black facial stripes were quite conspicuous; they looked plump and healthy, and I estimated that they were three to four weeks old. On being exposed they started to pull the hay round them (Illus. 18) and were soon buried again.

We went into the open shed where we had last seen the sow. There we found two big holes which she had dug in the earth floor about seven feet apart, and a third which she had abandoned on striking rock. Two large nest-shaped heaps of hay had previously been occupied, and a trail led about thirty yards into the barn from which she had brought the bedding. Evidently she had first tried to dig into the floor, then used a nest above ground in the open shed; but as this was probably too exposed she had made the final nest in which the cubs were born in the lean-to, using another abandoned nest there as a latrine. Taking a last look round, I shone the torch down one of the holes; two eyes reflected the beam. So that was where she had gone. The holes were connected, and she was lying midway between them. We left the shed, hoping she would return to her cubs.

Later that evening Jack Richards showed them to other friends. The sow had returned but, as they approached, she slipped quietly away, passing out of the shed and across the field in full view, without any sign of hurry. Next morning she was back again and curled up in the nest when the farmer looked in before breakfast; she did not move, but I wondered how long she would continue to put up with so much interference. When I visited the farm again that evening all was quiet in the shed, but I soon heard whickering from inside the heap of hay and saw it move a little. A careful examination revealed one cub but no sign of the other or of the sow in either the lean-to or the open shed. I concluded that she had removed one the previous night but did not have time to take both. The nearest set was about half a mile away in the canal bank, too far for the cub to walk, so presumably the sow had carried it in her mouth. If I were right she would return that evening for the second cub. The following morning, when I telephoned Jack Richards, he told me that the nest was empty.

The occasional badger nests recorded above ground have mostly been used for sleeping out, not for breeding. One found in Devon in June was in a dense rhododendron thicket through which I had to cut my way. Tracks led to a large heap of bedding where the badger had been sleeping regularly, presumably protected by the thick cover.

The only other breeding nest I have seen above ground was found a few years ago in a hedgerow on a peat moor near Westonzoyland, also in Somerset. The farmer, Sam Musgrave, heard whickering noises from a great heap of grass and reeds and wondered whether it was an otter's nest.

When he poked it with a stick, a sow badger shot through the roof and stood grunting at him before slowly retreating. Inside he found several tiny cubs with eyelids still fused, probably about a week old. I saw the nest a few days later, after the sow had taken the cubs away; the roof had gone, but it was still twenty-one inches high and more than two feet in diameter. The grasses and reeds seemed to be roughly woven together, making the thick walls quite firm and strong. It was understandable here that the sow could not dig into the peat, because the water table was nearly level with the surface; but why did the North Curry sow nest in a shed? Both appeared to be young animals, so perhaps they had been unable to establish themselves in more typical badger country.

Impatient dipper

Last Easter we were driving along the secondary road beside the River Mint in Westmorland when I caught sight of a dipper bowing on a boulder in mid-stream. He took no notice of the car, as he flew from one stone to another, then to the far bank where he picked up a brown beech leaf. With this he again flew from stone to stone in what I presumed to be a form of display. The hen now appeared from under the near bank and flew out to a mid-stream boulder. The excited cock, still bearing the leaf like a banner in his beak, flew up and down above her. She in her turn visited the far bank and vigorously attacked some dry bents which she cut down to size and arranged until she had a beakful protruding like whiskers. She dipped these in the water; then, part flying, part wading, part swimming, she crossed the river and disappeared under the near bank. – *Frank Bodman.* [Both male and female dippers build the nest, which is lined with dead leaves. Possibly the male was anticipating this stage; but it is more likely, as Dr Bodman suggests, that he was stimulating the female to build by carrying the leaf. Male finches, which do not help in building, also sometimes accompany their mates with material in their beaks. – *B. C.*]

Short back and sides

A donkey was reclining in the grass by Lough Rea, Co Galway, with a number of jackdaws in attendance. At first we thought they were delousing the animal, but further observation through binoculars showed that they were pulling hair from its neck, head and exposed flank. At one time there were as many as seven jackdaws at work, and the donkey just regarded them lazily through half-closed eyes. They were evidently gathering nest-lining in the easiest possible manner, because eventually they all made off towards some old trees and did not return. – *R. Harrison*

Fresh feathers

While watching the birds feeding in Lincoln's Inn Fields I saw a house sparrow hop up to a pigeon, pluck two feathers from the upper part of its leg and fly off with them to the eaves of the high buildings on the north side of the square. Beyond giving a slight hop, the pigeon took no notice and went on feeding. – *H. Stanley Jones*

Starlings, long established in New Zealand, have nested for many years in the holes of an old tree near our house. Just beyond it is a patch of soft ground where the hens scratch and peck all day, presenting to the rearward watcher a mass of soft fluffy lining feathers. A lightning dive, and a starling is away with a beak full of them. When the hens become sensitive and face the foe, the starlings appear to cooperate: one lands in front of the quarry as if to distract it, while the other swoops from behind. – *M. Ripley*

Temperature control?

A pair of long-tailed tits were investigating the site of their last year's nest on February 11th, 1961, and began carrying material to it three days later; but the first egg of a clutch of ten was not laid until March 22nd. When the young were about nine days old, we watched the parents bringing flies and small caterpillars; they spent from ten to fifteen seconds at the nest, feeding one and occasionally two gapes at a time. Then they would wait for a faecal sac to be extruded or remove a feather from the lining. Four days later only sacs, not feathers, were being taken; the brood of four fledged on May 17th. I believe that the removal of feathers may be a device to regulate the temperature of the nest as the young increase in size and the weather becomes warmer; no particular colour was selected. – *Helen Robbins*. [Megapodes and perhaps grebes regulate the temperature of their nests, so it is conceivable that similar behaviour has been evolved by titmice; but, when the drive to remove sacs is intense and none are immediately forthcoming, parents will often take away other objects from the nest or its vicinity, though these are generally whitish, like the sacs. – *B. C.*]

Ducklings' descent

As I stepped from my car by some wind-blown trees on the fringe of Morecambe Bay, I heard an agitated *quack-quack* and looked all round, even under the car. Then I saw a mallard duck take off from a horizontal branch at least thirty-feet up a pine tree. She flew around calling but soon alighted on the branch. Presently two ducklings fell down into the long grass and were followed by several more; all seemed to be unharmed. Still quacking, the mother took off again towards a pond three hundred yards away and flew to and fro, guiding them as they fought their difficult way to the water. – *Charles Sherdley*. [Although mallard quite often lay in old crows' and other nests high in trees, the descent of the brood is seldom observed. – *B. C.*]

Family fellowship

A pair of moorhens hatched three lots of eggs on our pond last year: seven in April, five in mid-June and five at the beginning of August. The first brood, eventually reduced to four, helped to feed the later ones until the third was about a fortnight old, when the parents drove them off; but the second brood, now of three, continued to feed the youngest, soon also reduced to three. As the fine weather continued, the water-level fell and the moorhens' usual roosting place became accessible to the farmyard cats. At the centre of the pond, which was bare of vegetation, stood the remains of a wire water-lily basket with the top two rows of holes above the water. On it, in less than two hours, the parent moorhens and two of the second brood built a substantial platform of waterweed nine inches high. When making the foundation one bird thrust a stem into the wire mesh, and a second pulled it through and into position; usually an adult worked with a juvenile at opposite sides. One parent then roosted on the platform with the three young, while the other went with the second brood into the tree overhead. – *Anne L. Cooper*

Seven to safety

We were gazing idly into my garden pool when a wood mouse came running towards us. It had been raining heavily, and under our feet was a large flat stone covering a cavity from which water was flowing into the pool. The mouse dived under the stone, to reappear quickly with a naked new-born baby the size of a shilling in her mouth. She raced across the path beside the pool, vanished under a bush and deposited the baby in a crevice on the rock face behind it.

After bringing out four more, the drenched and bedraggled creature ran to the lawn, where she preened and partly dried herself. But she was soon back, her task now easier because the water had stopped running. Altogether, in about three minutes, she removed seven babies from the old nest. It had been built during dry weather in a spot which appeared to be safe. – *Fred J. Chapple*

WEASEL RESCUE by Ann Blakiston

As I was sitting by the river one evening in early summer, I heard shrill squeaking which seemed to come from a clump of docks on the opposite bank. Suddenly one weasel, then another and finally two babies came running from under a leaf and went scampering and calling along the bank, so I followed them on my side. Just ahead the river turned at right angles through hatches and flowed on through meadows to the withy beds. The leading weasel, which I took to be the female, made several little rushes to the water's edge and then ran on again. After she had done this three times I saw a third baby in the water being swept along towards the hatches, to which the other weasels now ran. The mother left the father to guide the two babies across, while she climbed down the centre post until she was at water-level. The current was quite fast and, as the baby was swept up to her, she grabbed it by the scruff of the neck, holding it as high out of the water as she could. Swimming back to the post, she caught it with her front paws and slowly hauled herself and the little one out of the water. The father had deposited the other two on the bank and was running up and down the hatch plank, squeaking.

When the mother had got halfway up she slipped and fell back into the river, and the whole agonizing performance started again. Once more she fell. I was desperate to help but dared not move in case I frightened her into dropping the baby. At the third attempt she got past the slippery patch which had been her undoing, but was too tired to climb any higher. The father weasel then suddenly climbed down the post head first, got hold of her by the scruff and dragged her and the baby up inch by inch, the mother helping by pushing with her hind feet. At last they were safely on the hatch plank where they lay panting, the mother still holding the baby; after about a minute they scrambled to their feet and scuttled across the plank to join the other little ones. The squeaking started again, and all five trooped past me in single file down the meadow.

CHILD CARE by Marjorie Holland

Although it was fine, the Thames at Maidenhead was running high as the family of swans from the eyot set off downstream – cob, pen and six cygnets about a week old. Late in the evening I saw them returning, four on the mother's back, close to our shore where there is always a slight ebb. They crossed the twenty yards of rushing open water to the islet, where their troubles began. Time after time the pen would half-fly, half-scramble up the steep bank, trying to coax her babies to follow. The cob remained on guard in the water; if one did drift away, it bumped into him. Then he would turn and, protecting it with his body, shepherd it into position for another attempt.

When it became obvious they were tiring and could not land unaided, my daughter and two friends changed into swimsuits and crossed over in their canoe. After several vain efforts my daughter suggested paddling among the brood while she tried to grab them. She got one in each hand at her first attempt and put them on the islet. The cob landed and hustled the bedraggled cygnets over to the nest, where the pen settled down with them. But when five were safe the parents took no further interest and the sixth, without the cob's body to protect it, drifted some yards away, cheeping in distress. Once it was in the main current, rescue would be impossible. Stretching to the limit, the paddler put his blade under the cygnet and shoved it towards my daughter, who scooped it on land. But it was too exhausted or scared to move; so she climbed ashore and carried it over to the nest, where it scuttled under the pen. The cob, standing guard, made no attempt to attack her.

Sitting it out

Five pairs of lapwings bred near my cottage in Mid Yell last April, so I put up a hide fifty feet from one of the nests and watched the bird return to her eggs safely. Next morning I awoke to a small snowdrift on the bedspread, and four inches of snow lay outside. About eleven o'clock the sun came out and, fearing the worst, I visited the lapwings. All five were on their nests, some with only head and tail showing. The snow melted quickly, so by early afternoon I felt it was safe to enter my hide and, though I did not like to move it any nearer to the nest, I was able to take a series of photographs (Illus. 19). All five broods hatched out successfully.
– *Bobby Tulloch*

THROUGH THE STORM by Phillip Glasier

The bird photographed in Illus. 7 and 8 laid only one egg, instead of the more usual two, in a mere scrape between drills of barley on a farm in Wiltshire. Some ten minutes after a friend had left me in the hide he had erected nearby, the stone curlew flew round, pitched on the ground some way off and approached in short little runs, stopping at intervals to listen or to pick at something on the ground. She settled on the egg and sat fluffed out with her head well up. I had been in the hide for almost two hours and taken several photographs when I noticed a slight movement under her left wing, and there was the chick's head peeping out. The hen seemed to be so unconcerned that I hoped to photograph the chick when she got up, but within a very short space of time the skies darkened and we were in the middle of a torrential thunderstorm. The hen sat tight-feathered, so that the water would stream off her back, and at the height of the storm her head was right down on the ground. As soon as the rain eased off, I hastily got my camera into position again and took the lower picture of the drenched and be-draggled bird. A few minutes later the sun came out, and she shook herself and fluffed out her feathers. She was soon sitting up quite dry again, looking a very different bird from the one crouching flat in the downpour.

Shrike curiosity

The photograph (Illus. 20) shows a red-backed shrike's nest adjoining and intertwined with the previous year's, so that the pair looked like a figure of eight. The old nest was well preserved, and the young hopped from one to the other, apparently quite indifferent as to which they occupied. J. H. Owen, the shrike authority, told me that he had known only one other such nest. To obtain the cock bird's picture I played a trick on him. By placing a penny flat in the palm of one hand and striking it sharply with the edge of another a very fair imitation of the shrike's *chik*, *chak* is obtained. When I tried this, the cock came down at once to investigate the intruder. – *R. P. Gait*. [Actually, the bird in the photograph, with a scaly breast and rather faint 'mask', looks like a hen shrike in partial cock's plumage, a quite frequent occurrence. – *B. C.*]

Multiple nests

A ladder hung against a wall is the classic situation for multiple nest-building, but it can also occur wherever man's constructions present a row of identical sites. Usually, the bird lays only in one nest, abandoning the others in various stages of completeness; sometimes two clutches are begun, but only one completed. Blackbirds, song thrushes (Illus.

22. Our largest and rarest bush-cricket

23. (*above left*) Swallowtail caterpillars on carrot; 24. (*above right*) pairing at Wicken Fen

25. Adult laying eggs on fennel

26. Peppered moth and its black variety on sooty bark

27. On lichen-covered bark

28. The thrush that couldn't make up its mind

29. Hedgehog and parasites

30. Coypu with its head on the water

31. In profile

32. Red-necked phalaropes on a Shetland loch

33. Owl's bath: on the brink

34. Owl enjoying the dip

35. Before the dust bath

36. After the dust bath

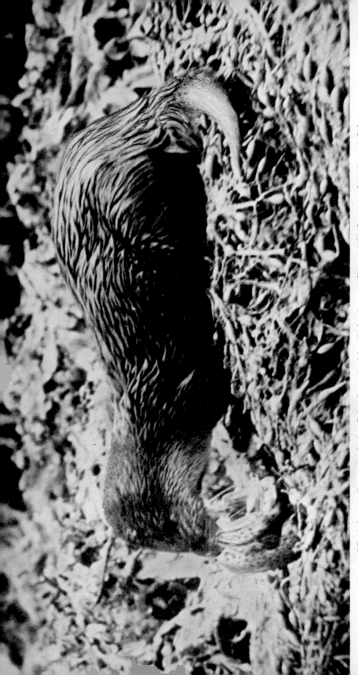

37. (above) Otter and dogfish alerted; 38. (below) wres·ling with the tough-skinned fish

39. Otter pops up

40. The Blind Lady and her calf

28) and robins are the species most often concerned, but a blue tit in the Forest of Dean, Gloucestershire, completed clutches in two compartments of a triple nest-box and tried unsuccessfully to incubate both. – *B. C.*

Two cuckoos

Walking in the fields near my home I heard a bird's call repeated in the same key over and over again. After some searching I found that it came from a fully grown cuckoo on a wire fence. While I stood there it was joined by another, which immediately set up the same continuous call. This went on for almost five minutes, until quite suddenly a meadow pipit arrived and hopped on the back of one cuckoo, which turned its head to take a morsel from the small bird's beak. The meadow pipit flew off again, and once more the air was filled with the cuckoos' calls. The second was fed in exactly the same manner, and feeding continued in strict rotation by what I thought was one bird until two meadow pipits arrived together. For two or three days I saw the ritual repeated at the same spot. One day I watched it for more than an hour and a half. – *P. J. Lennon.* [The meadow pipit is a common fosterer of cuckoos, and it is quite possible for two cuckoos to lay in the same nest. It is much rarer for the two young ones, if they hatch, to reach a *modus vivendi*; but that has happened and might be the explanation here. It is about equally possible that the two young cuckoos were hatched in neighbouring nests and were being fed by one parent from each. – *B. C.*]

FOSTERED BY FLYCATCHERS by Doris Theak

One summer a pair of spotted flycatchers reared a cuckoo in their nest, which was in a recess about twelve feet from the ground in a stone fruit-garden wall. For a week before it flew they came with beakfuls of flies from 4.30 AM until

10 PM; four minutes was the longest interval recorded between visits. Thus they were at work for seventeen and a half hours at a stretch, and they continued to feed the cuckoo for several days after it left the nest early on June 25th. Its appetite seemed to be completely satisfied because, though it swallowed every beakful of food with alacrity, I did not hear the hunger cry, and I was within easy hearing range. This was the reverse of the behaviour of young cuckoos which I had watched being fed by robins and meadow pipits; they called incessantly. Silence may have been an idiosyncrasy of the flycatchers' cuckoo, but it might be hard, particularly in dry weather, for robins or pipits to find enough worms, caterpillars or beetle grubs, while the flycatchers could collect flies rapidly during those hot midsummer days. After the cuckoo had fledged, the flycatchers reared in the same nest a brood of their own which flew early in August. A second young cuckoo was hatched in a flycatcher's nest in another garden half a mile away, and it seems probable that the egg was laid by the same female. [It is comparatively rare for a cuckoo to fledge from a spotted flycatcher's nest; the egg is usually rejected. As the note suggests, feeding is no problem in sunny weather, but it might be difficult in a wet summer when insects would be reluctant to fly. – *B. C.*]

LAID OUT by Florence E. Pettit

On May 11th a cuckoo glided silently past my window in the early morning – evidently a female looking for somewhere to lay an egg. Two days later I left my desk momentarily about 9.30 AM and returned to see some small grey feathers stuck to the window above my typewriter. On the concrete path outside the glazed door sat a cuckoo whose rufous-tinted breast indicated a female. Her eyes and beak opened and shut slowly several times, and I concluded that she had crashed into the window-pane. Apart from loss of feathers she seemed undamaged, so I put an open-slatted plant-box over her to keep off the sun. After about an hour, during which she sat immobile like a china ornament, I started to stroke the top of her head with a soft dry paintbrush through the slats. She blinked, as though I had broken a spell, and moved normally, turning to find a way out of the box. Then I saw the egg that had been under her. I lifted the box, and she flew to the branch of a flowering viburnum, where she lay draped for another hour and a half, ignoring the protests of a blackbird. Soon after midday she rose over the beech hedge and away. She had been within a few yards of a suitable home for her egg – a dunnock's nest behind screening golden balls of *Kerria japonica*.

Part Seven

Scale-Wings

Introduction

Butterflies and moths do not provide subjects for many
short notes in *The Countryman*, so I have made a choice of
three articles which have appeared in recent years, all by
specialists in this most popular and colourful order of insects. The working of natural selection as demonstrated for
the two forms of the peppered moth by Bernard Kettlewell
and so well exemplified here in two of Michael Tweedie's
photographs is one of the outstanding stories of genetical
field research since the war. Dr Kettlewell described it himself in the magazine *Discovery* for December 1955.

Two short notes concern an unusual pet, and the delight
which the most spectacular of British butterflies can give to
busy men on a day off in the New Forest.

SWALLOWTAILS FOR WICKEN
by Brian Gardiner

When I first set foot on Wicken Fen it was as a young and
enthusiastic collector of butterflies, and I can still remember
the thrill of seeing and catching my first swallowtail. In
those days I used to cycle to the fen along the towpath of the
Cam and cross the river by the ferry run by the landlord of
the No Hurry inn—now, alas, no more. When one was
exhausted from chasing butterflies refreshment could be had
near the entrance to the fen at the Black Horse, now also gone.
It was here that I heard tales of 'enormous moths, big as

birds', which were driven off the fen by the great fire in 1926. Like the fish that got away, a moth grows in the telling thereof.

Fifteen or more years ago, on a walk down one of the rides, I could generally see several swallowtails in flight, and a diligent search along the edges of the ride would produce a dozen or more caterpillars, for the milk parsley on which they fed occurred every few yards. There was a rule in those days that no one might catch more than six specimens a year. Since I found the flying butterflies difficult to catch and then had no facilities for breeding caterpillars, my total bag was two, which I still have. Today I breed them by the hundred. When I release them over the fen and watch them fly off in search of food, two or more often choosing the same flower, I get as much pleasure as I did from that first sight of them nearly twenty years ago.

There seems little doubt that by 1950 the swallowtail was extinct on the fen after being isolated there for nearly a century. Like much in ecology, the cause is difficult to ascertain. In my view it was mainly the result of overgrowth by the wrong type of vegetation. When I first visited Wicken there were open tracts of sedge, and the food plant was plentiful. Looking down over the fen two years ago from the top of the new bird hide I was reminded of an aerial photograph taken over forest. Apart from a few laboriously hand-cleared areas it is now mainly a thicket of buckthorn, birch and alder. In a walk of more than two miles recently I was able to find only three plants of the milk parsley on which the swallowtail caterpillars feed; but it is becoming re-established in the newly cleared area near the entrance to the fen.

It may also be that the floods of 1947 contributed to the demise of the species. It certainly survived the great fire of 1926, which had, if anything, a beneficial effect. In a burnt area, the first to reappear are the smaller plants on whose flowers the butterflies feed. Indeed there is some evidence that, as these became overgrown by scrub, the butterflies left in search of food, many of them perhaps before they had had a chance to lay their eggs. The wartime drainage and ploughing of Adventurer's Fen, now happily being restored to marsh conditions, also nearly halved the area available to them.

When I was given the opportunity of breeding swallow-tails to restock the fen I accepted the challenge. It is often easy to breed a few butterflies for a year or two, but the production of large numbers over an extended period brings special problems. Chief among them is the danger of an outbreak of some bacterial or viral disease, and there is also the possibility of deleterious effects of in-breeding. I had already had experience of breeding some hundreds of thouss-sands of cabbage whites, so it seemed reasonable to apply the same techniques to the swallowtails; but they proved to be rather more difficult. The cabbage whites are ready to feed on artificial flowers made of painted Perspex and containing a small glass tube of honey solution. The swallow-tails completely ignore these, so I have to feed them on cut blooms, using chiefly valerian and sweet williams, which have to be changed at least every other day. Towards the end of the swallowtail season it is quite a job to find enough of them.

When the sun is shining the butterflies readily mate and lay their eggs. Now it is a curious fact that, while in the wild swallowtail caterpillars usually feed on two marshland plants, in captivity they will take to quite a wide variety. For egg-laying (Illus. 25) I use either fennel or wild carrot, as well as one of the natural foods, milk parsley. I sow them in pots in the autumn and grow them in a greenhouse. For the larger caterpillars I use the leaves of the garden carrot (Illus. 23). Neither the egg-laying butterflies nor the feeding cater-pillars seem to show any preference for their natural food over the others, and it is rather difficult to explain why this lovely butterfly should be confined to the Norfolk Broads and the Fens, for both fennel and carrot are widespread, and they are used as food on the Continent; but then the English swallowtail is a distinct sub-species.

When dealing with large numbers of caterpillars I have to make sure they do not become overcrowded, for this might well lead to an outbreak of disease. In the five years I have been breeding them I have so far managed to avoid disaster. Fortunately, insect diseases are specific, for when I started swallowtails I was losing about ninety per cent of my cabbage whites to a virus.

A cage full of swallowtail caterpillars has a pleasant pineapple smell, given off by the reddish *osmaterium* situated just behind the head. This odour seems to be a defensive mechanism which comes into play when the caterpillars are disturbed. Apparently, it is effective, for I have had no trouble from parasites with the caterpillars. The chrysalises, on the other hand, have been heavily attacked. To protect them I have had to use double-walled cages or to place one or more small cages inside larger ones. Even so, a few are attacked, and I must examine them carefully every week, destroying those that have been 'stung'. One of the habits of the parasite, known as *Pteromalus puparum*, is to sit beside the caterpillar until it sheds its skin and becomes a chrysalis. The parasite then proceeds to lay eggs through the soft skin of its helpless victim. When, as often happens in my greenhouse, the parasite has a choice, it prefers swallowtails to cabbage whites, which at other times it will heavily attack.

After several hundred eggs have been laid I collect up the adult butterflies for release on Wicken Fen. As a fine day is required for this operation, I sometimes have to keep them for several days in a cool dark room. They will live for two or three weeks without food in these conditions, and I believe that, barring accidents, butterflies live much longer than is generally supposed. For travelling I use a muslin cage a foot square, with a sliding glass front. When this contains a hundred swallowtails it is a fascinating sight because, unlike some butterflies, they do not flutter about madly, but rhythmically open and close their wings when disturbed. I remember one gloriously hot summer day when the sight so intrigued a policeman doing a traffic check that the question of my licence, left at home in my jacket, was quietly ignored.

On reaching Wicken I walk along the droves and every few hundred yards release some of the butterflies. They are hungry after their journey and almost at once seek out marsh thistle flowers and start to feed. They are still quite tame and can easily be recaptured without the aid of a net. I once spent a few enjoyable hours posing these butterflies to be photographed (Illus. 24). They turned out to be good actors, and I thought how different it all was from my first encounter when I spent hours fruitlessly chasing them.

BUTTERFLIES FROM SPAIN
by L. Hugh Newman

If butterfly migration were suddenly to cease, our summer scene would be much the poorer. We would have no red admirals in our gardens, no painted ladies flying across the countryside, no clouded yellows on lucerne fields, and there would certainly be no Bath whites to excite entomologists. It has been known for a long time that all these butterflies cross the Channel in varying numbers every year, with the exception of the Bath white, which comes as far north as Britain only on very rare occasions. The last immigration occurred in the summer of 1945 when this rather insignificant butterfly landed in unprecedented numbers at points all along the South Coast, the main influx being to Devon and Cornwall. What many people do not realize is that the migrants are continuously brooded; they have no long resting season. Exactly where they spend the winter has not been at all clear. The South of France has often been mentioned, but when I toured the Côte d'Azur in the early spring three years ago the only migrant I saw was a single clouded yellow. There were no painted ladies and red admirals, nor could I find their larvae.

Last year in early February I spent some time in southern Spain on the Costa del Sol between Gibraltar and Malaga. While I had expected to see a few butterflies on the wing, I was quite unprepared for the enormous numbers I met on my travels. Everywhere I found clouded yellows, Bath whites and red admirals, and it was obvious from their behaviour that they were not just passing on a northward journey. They were undoubtedly resident. Some were quite worn, having been on the wing for a considerable time; others had only just emerged and I saw many on their maiden flights. I watched their courtship, feeding and egg-laying on herbage which was green and fresh. Soon I was convinced that I had discovered the real winter home of these summer visitors to our shores.

The north coast of Africa has been generally accepted as the starting point of butterfly migrations, but I am quite sure that the insects I saw had never crossed the Straits of Gibraltar. The painted lady, I admit, may well come from Africa; certainly I did not see any in Spain during February. Two eye-witness accounts, one over a century ago, the other dating from the last war, each describing the start of a northward movement of painted ladies from Africa, lend support to the belief that this particular traveller begins its annual migration outside Europe.

The most interesting observation I made during my Spanish excursion was brought home to me forcibly the moment I stepped from the plane on to the long landing strip at Gibraltar. It was a beautifully warm and sunny day – I remember the soft feel of the air and the summer scents after wintry Britain – and everywhere I saw large cabbage whites floating like huge snowflakes in the air. My first thought was that I had landed in the middle of a large migration, but as the days passed I realized my error. Wherever I went along the coast there were cabbage whites, quantities of them and obviously local residents. There were plenty of wild and cultivated brassicas on which to lay their eggs. Many were just coming into flower but would last long enough to feed the next generation of caterpillars. Thereafter, owing to the summer heat and drought, there would be little food, and the butterflies would almost certainly start a northward movement.

The migration of the large white has always been something of a mystery. No one has stated with any authority where the large swarms which usually reach England in

August originate. The standard authority on butterfly migration speaks of Scandinavia and the Baltic area as their breeding ground, from which they are supposed to fly in a south-westerly direction before they cross the North Sea to England. I have never accepted this theory. Somehow it did not seem convincing; and the Scandinavian farmers do not grow cabbage crops on a large scale. I am now of the opinion that the large white moves up from southern Spain by easy stages in a north-easterly direction, helped by prevailing winds which bring it to the French Riviera, through the mountain passes skirting Germany and then into the Netherlands. By the time the third generation of the year has reached this part of Europe, about midsummer, the numbers have nearly reached dispersal proportions. From there, with the urge to migrate still in their blood, their descendants spill over into Britain. The usual route is across the Essex marshes, but the path often extends farther south; I can remember one August Bank Holiday when hundreds of thousands of large whites descended on Broadstairs. They were tired and hungry after their long flight and settled immediately on the flowers along the front and even on the scarlet uniforms of bandsmen, mistaking these no doubt for some exotic bloom.

Our native large white is essentially a double-brooded insect and winters in the chrysalis stage; but it can be induced to breed continuously under artificial conditions in Britain, and in a warmer climate that is undoubtedly what happens. The Baltic theory does not seem to take into account the fact that spring comes later there than in England, so that in Baltic countries the large white would not have time to build up the populations necessary to trigger off a migration. In southern Spain, on the other hand, it is certainly warm enough for them to breed all the year round, and they have ample opportunity to multiply during the movement across Europe.

Only fairly recently has Spain become a popular holiday country, and I doubt whether any inquiring traveller with special knowledge of butterflies has visited this part of the coast in winter. Most collectors take their holidays in summer, when the migrants would have left for more temperate northern climates. This is probably why the Costa del Sol has not yet been recognized as an important butterfly nursery. Trained observers send regular reports to the Insect Immigration Committee from both the Atlantic and Mediterranean coasts of southern France, but I am not aware that anybody has recorded movements in southern Spain. I wish I could spend a whole winter there and keep detailed records of the dates when the various butterflies appear and of any northward flights. I believe the results would throw a great deal of new light on this still rather obscure subject of butterfly migration and help to answer many of the questions that still puzzle us.

BUTTERFLY IN THE HOUSE by Ida Lumley

At Newport, Isle of Wight, my husband's study became unusually warm one early winter's day, and a peacock butterfly emerged from hibernation to drink water that had condensed on a window. I transferred it to a cooler room, where it evidently found a dark corner. We did not see it again for six weeks, until one morning I heard a tiny rustling noise and saw the butterfly on a curtain in the sunlight, gently moving its wings to and fro. I soaked a weak mixture of honey and water on a small sponge. After a great deal of persuasion the butterfly started to feed – and continued for ten minutes without stopping. After that it woke up at intervals of between two or three days and a week or ten days, according to the weather. Each time I fed it, and it became more used to the procedure. Eventually I needed only to hold my finger near it, and it would at once tap with a leg before walking on to my hand; yet it refused to come to my husband or any other member of the family. As the weather became warmer the peacock spent more time awake and would fly about the room, returning to me. It began to beat its wings against the windows, so I took it into the garden, where it stayed for about half an hour before flying indoors. This happened again when I let it out through the french window. The third time it did not return.

PURPLE HIGHFLYER by A. M. G. Campbell

One of the greatest thrills for the butterfly enthusiast is to see his first purple emperor. This was certainly true last July for three medical men on a visit to a locality where purple emperors have been known for some years. In three days we saw several on the wing. A female remained twenty feet up on a sallow for about half an hour to give us excellent views. Two males fluttered down almost to ground level to inspect a motor-scooter and car; the vehicles were reflecting the sun's rays, and either this or their bright colours attracted the butterflies, which are curious about unusual objects as well as smells. One was found dead in a farmer's car. Others had visited the midden in his yard several times during the month, chiefly in the mornings and evenings. We also watched a male for more than two hours, during which it kept to a rhythm of behaviour, retiring to sit on the same leaf after short flights of five minutes each. It was about until after 6 PM, flying in dull spells as well as in sunlight. The flight consists of powerful and rapid wing movements, followed by prolonged glides, and justifies the old name of purple highflyer. As its food plant, sallow, is unlikely to be contaminated by insecticides or grazed by stock, the species has reasonably good prospects in those parts of England where it is still found.

MOTH CAMOUFLAGE by M. W. F. Tweedie

Moths are eminently edible creatures. A few, like the burnets and the cinnabar, are protected by unpleasant flavours and concomitant lurid colours, but most are welcome items on the insectivorous animal's bill of fare: generally, if a bird finds a moth it will eat it.

The majority of moths fly by night and are inactive during the day. Since birds hunt by sight, the moths must hide; and most of them conceal themselves in thick vegetation. Agrotids such as the heart-and-dart and yellow underwing creep down among the grass stems. A large number, including the carpets, waves and others of the Geometrid family, hide among foliage. Some, however, choose to sit in the open, usually on the trunks and branches of trees; and it is these which afford impressive examples of natural camouflage, necessary for their survival.

The common shark is a moth I find by searching systematically on fence posts. I believe it has a preference for the relatively smooth surface that bare wood affords; and I have not come across one on the bark of a tree, where it would certainly be more conspicuous. Its finely streaked wings simulate the grain of naked wood far better than they would the irregular pattern which bark and lichen present.

The Clifden nonpareil is one of our largest moths and a great rarity; its broad grey forewings are marbled beautifully in accord with lichened trunks. Distributed through woodland at about one to the acre, these moths would take some finding. The lovely merveille du jour, being marked with black, white and pale green, is an even more finished simulator of lichen. It is common in autumn wherever there are oak trees.

The most exciting and illuminating known example of moth camouflage is shown in Illus. 26 and 27. Before 1850 the peppered moth was represented in British collections by the black-speckled white form which gave the species its name. In that year a black variety was taken in Manchester, where it soon became common; and during the past century this black peppered moth, var. *carbonaria*, has largely replaced the typical form (*betularia*) in and around our large towns. It is a feature of the modern urban environment that the lichen on tree trunks is killed by the smoke-contaminated air; and the bark itself is blackened by deposited soot. In such surroundings the typical form of the moth, sitting on a tree, would be conspicuous at fifty yards, whereas *carbonaria* is invisible at five. Take the two out into the country where the trees are clean and covered with healthy lichen, and the case is dramatically reversed. The photographs do not, of course, represent pairs of moths found naturally in convenient juxtaposition. They were posed to illustrate the principle of industrial melanism, as this phenomenon is called. Look fixedly at one of the photographs and you will see three moths, not four; the 'tail' of your eye will not resolve the camouflaged individual. Then look fixedly at the other, and again only three moths will catch your eye.

Industrial melanism was first noticed in the peppered moth but has occurred also in several other species. It is one of the very few evolutionary changes whose progress has been observed.

Part Eight

Mainly Mammals

Introduction

Another short Part takes in some articles and notes which cannot easily be classified elsewhere. The articles deal with an ancient and a recent introduction; John Buxton's description and photographs of the coypu, written about ten years ago, are already of some historic interest. He is now well-known for his films in Anglia's *Survival* series.

Gwendolen Barnard's note is one of three published recently about dog-fox friendships in the wild; and Chris Mylne, whose films have also been seen by millions on television and by thousands at meetings organized by the Royal Society for the Protection of Birds or the National Trust for Scotland, applies a technique developed at the bird observatories to the removal from a hedgehog of the numerous parasites, which are apparently its normal companions.

FALLOW DEER by P. H. Carne

Few English counties are without wild fallow deer, and they are also at large in many parts of Scotland and Ireland. Their relative rarity in Wales, where there is suitable country in plenty, is due solely to the small number of parks from which deer have escaped. The border counties of Herefordshire and Shropshire have many such parks, and wild fallow are plentiful around them. This may also be said of almost every locality where escaped animals have access to extensive woodland. At one time, comparatively few gained their freedom, and on well-keepered estates these were soon shot; but the decline in the number of keepers and the neglect of fences during and after two world wars led to widespread and successful breakouts. During the 1939–45 war at least one owner deliberately released his herd in anticipation of a ploughing-up order, and several hundred members of it still occur wild in the surrounding woods. The legal definition of park deer as 'wild animals under restraint' is especially apt for fallow. Red deer which find their way through a hole in a park fence often roam more or less aimlessly until they are driven back. Fallow, on the other hand, soon discard their tameness, which at best is only a thin veneer.

In parks these deer tend to gather in large assemblies in the open. When they escape they become woodland dwellers and small parties are the rule: only in late winter and early spring, or for a short time during the rutting season, are large herds likely to be seen. In March, April and early May I have seen up to twenty-five does and yearlings together, and troops of upwards of twenty bucks are not uncommon in the New Forest for a few weeks before they shed their antlers in May. The bucks then disperse in ones and twos looking, and

perhaps feeling, very sorry for themselves for several weeks.
Not until the new antlers are well grown do they resume a
loosely organized social life with their own kind. Apart from
a brief interval at fawning the does remain together throughout the year. Twin foetuses have been observed once or twice
in the gralloch of New Forest does, but plural births are
extremely rare.

Having cleaned or 'burnished' their antlers by late August
or early September, some of the older bucks begin to follow
the does about four weeks later. Belling or 'groaning' begins
about October 5th and normally reaches its peak between the

15th and 20th of the month. Unlike the intermittent roar of
red deer, the pig-like grunt of a fallow buck maintains a
constant rhythm when a harem of appreciable dimensions is
in dispute. The master buck relies mainly on vocal prowess,
coupled with hurryings and scurryings to and fro among his
charges, to restrain the more wayward among them. Lesser
bucks patrol the outer fringe of the harem zone and miss no
chance of stolen amours when the master's back is turned,
though a snorting charge is usually enough to deter his more
importunate rivals. Duelling is generally a formal thrust-and-
parry affair that results in little damage, but fatalities do

occasionally occur. Among few recorded instances is one of two New Forest bucks found dead from starvation with antlers interlocked. Younger bucks take what is left when the old ones are 'run out', and by early November the rut is virtually over. Sometimes bucks and does consort together through the winter, but in my experience the sexes are more often segregated.

There is still much to be learnt about the migratory and territorial habits of wild fallow. Even in so extensive an area as the New Forest individual bucks and does appear to remain in particular zones from which they rarely, if ever, stray of their own accord, save during the rut. Certain woods, for example, are inhabited by white deer whose daily and seasonal movements can be easily plotted as they circulate within quite well-defined and closely limited territories. There is also evidence that some woods are favoured mainly by does, others by bucks. Occasionally fallow deer migrate considerable distances for no apparent reason except a desire for change, and they then follow regular routes.

Railways play a varying role in restricting the movements of these deer. They cross some lines with impunity and ob-

vious regularity: between Powerstock and Toller in West Dorset a recognized fallow pass traverses the Maiden Newton–Bridport branch line. In the New Forest, on the other hand, it is almost unknown for fallow to cross the Southampton–Bournemouth railway except at one point near Lyndhurst Road station. Elsewhere they are hardly ever seen to the south of it. Open country also deters free movement of wild fallow. At night they will regularly trespass on farmland, but they rarely wander more than a mile or so from the cover of trees. Unlike roe, they are rather reluctant pioneers of unknown territory. Thus, escaped animals tend to establish isolated woodland colonies fairly close to the parks whence they came, the extent of their habitat varying hardly at all from year to year, although at rutting time bucks may travel twenty miles or more from their original abodes. It is not at all unusual for such bucks to enter deer parks and there unwittingly conclude their career as feral creatures.

BUSBY AND FRIEND by Gwendolen Barnard

Some years ago I used to walk daily through a Hertford-shire wood with two labradors. In March I found a fox's earth there and had difficulty in preventing one of the re-trievers, called Busby, from investigating it. A week or so later he returned from a walk on his own, filthy and smelling strongly of fox. For some weeks afterwards he refused to pass the earth, sitting down and waiting while I walked on with the other dog; and when I went back one day, I saw him intent on smelling a fox cub. In June my husband took both dogs into the wood and lost Busby, who reappeared trotting towards him with a young vixen. The dog was often away in the following weeks, coming back caked in sandy mud; he had obviously been rabbiting. One day, waiting upwind of his hunting ground, I saw him arrive with a fox and begin to dig furiously while it lay a few yards away. His disappearances continued, but I did not see the two together again. The following February, as the labradors and I were watching the hunt draw a field of kale in the valley below, a fox came straight towards us, stopped within three yards, sat down and looked at Busby for about half a minute before vanishing across country; by this time hounds were in full cry. Busby did not smell of fox again, and his rabbiting expeditions became fewer.

FIELD MOUSE'S TEA-PARTY by Alan Langman

One autumn afternoon, when we were picnicking by the osier beds of the River Parrett, a female long-tailed field mouse slowly emerged from the grass and crept towards the teacloth. At its edge she stopped, sat slightly raised on her haunches and fixed us with beady black eyes; her whiskers seemed to twitch in anticipation. Somehow I felt this had happened to her many times. I placed before her a small piece of cheese, which she immediately picked up and ate. She sat up again and waited for more; biscuit and fruit-cake crumbs went down with relish. When I tried stroking her head and back, she showed no sigh of fear or resentment; but when I lifted her tail, she spun round in annoyance. She consumed a great quantity of biscuit, cheese and cake for her size, though she was fairly corpulent with a family well on the way; but she knew her limits. Suddenly she refused any more crumbs, hunched up into a compact ball and went to sleep. We drank our last cups of tea, watching the sun set behind distant hills, and still our small companion slept on. The noise of packing awoke her and she began a vigorous and thorough grooming, with special attention to the place where I had touched her. Her whiskers came in for the final touches. Then, as suddenly as she had come, she vanished.

SHEDDING THE LOAD by C. K. Mylne

Hedgehogs are notoriously verminous and, when we found one which had wandered on to the precincts of the King's

Buildings at Edinburgh University looking dirtier, smelling more unpleasantly and obviously carrying more fleas than usual, we decided to investigate its menagerie carefully. Both ears were so full of bloated ticks that we doubted whether the animal could have heard anything. We removed them, but to pick off the fleas individually was a waste of time: the first twenty took us as many minutes. So we used the delousing technique evolved at the Fair Isle Bird Observatory. The apparatus consists of a plastic cylinder, open at both ends, into which chloroform is pumped while the body of the bird is held suspended inside, its head protruding through a hole in a plastic apron.

We had to improvise with a polythene bag and soon found that it was much easier to hold a bird with its head outside the vapour bath than a hedgehog. Our chief difficulty was to persuade it to uncurl enough for us to put its wet little nose through the hole we had cut in a corner of the bag. When patience had achieved this, we inserted a tube and pumped air through the chloroform bottle into the bag. The fleas started to drop off at once. We shook our patient to loosen them, and the total was obviously impressive when we released it from the cleansing; but there were still live fleas on it. We picked off many, then put it on a wire-mesh tray and rolled it gently to and fro while more and yet more dropped off; most of them were dead, but some were still kicking (Illus. 29). The grand total was thirty-one large ticks and five hundred and eighty-four fleas.

On closer examination we discovered there were also hundreds of tiny ticks embedded in the hedgehog's limbs and the skin of its underside. Many were too small to be pulled off intact, but those we could remove seemed to have died from the chloroform, so the rest would presumably drop off in time. Many of the fleas were gravid females, yellow-brown and much larger than the dark brown males. Dr W. Sinclair of the Zoology Department was delighted at the haul of specimens for his research; but the hedgehog may have been less pleased with the thorough treatment, because there is some evidence that the removal of so many parasites at one time leaves an overdose of antibodies in the bloodstream.

Brock gated

Surprised by our headlights, the badger which visits us from the Highmeadow Woods above the Wye tried to escape through a latched rustic wicket-gate. He shoved his head between the round spars and, by pushing hard, forced his body through to the ribs. But he soon lost what little grip he had with his claws and, as the gate was shut on a high step, his head and shoulders hung helplessly while his hindquarters were jammed. We opened the gate gently and suspended him; he wriggled sideways, we pushed behind, and with a cough and a grunt he was free. For a few moments he sat where he had landed, then got to his feet, shook himself vigorously, scratched his neck and trotted off into the night. – *Hilda N. Taylor*

COYPUS IN NORFOLK by J. J. Buxton

Various creatures now resident in the British Isles have been introduced at one time or another. Many of them we take for granted: grey squirrels (in some areas) and pheasants, for example, are part of the everyday scene. Rabbits, too, have been here a very long time. We now have a new animal. the coypu, which in certain districts is obviously going to become one of our commonest.

The ancestors of the coypus which now infest many of our slow running rivers, particularly in East Anglia, lived in the vast swamps of Brazil. Specimens were brought to Europe between the wars to produce the fur known as nutria. The coypu is not only a pest but has an unattractive appearance. It is a rodent with some of the rat's most unpleasant features (Illus. 31), magnified by its size, and large, bright orange front teeth. Its long, rather coarse outer fur is the colour of a brown rat's; but the more mole-like inner fur so prized by the ladies is dark chocolate and very soft.

When swimming (Illus. 30), a coypu looks rather like an otter, but with less back showing. The tail is much thinner than an otter's and does not appear so much like a long plank of wood on the surface. Only the hind feet are webbed, and they are considerably bigger than the front ones. Long sharp claws exist on all four feet, enabling their possessors to dig powerfully. The coypu affords a good example of the way Nature suits an animal to her environment. The mammary glands are situated along the back, so that the young can feed while their mother is in the water. It is a quaint sight to see three or four floating at right angles to her, as she lies with only the top of her back and head above the surface.

Being strictly vegetarian, coypus do not scavenge as much as rats; but they have taken a great liking to various farm crops, of which sugar-beet and mangolds probably top the list. Kale fields near the Norfolk dykes attract them, and growing corn is often badly damaged. They eat grass, particularly in frost, when nothing better offers; but here the worst damage from the farmer's point of view is that caused by their numerous tracks: coypus are expert at causing a hay crop to lodge, so that it becomes difficult to cut. Their commonest foods include the roots of water dock, lesser bulrush, candle rush, sedge and other plants abundant round the Broads. The young shoots of reeds also receive some attention. Tame or enclosed coypus are quite happy with bread and biscuits, and also appreciate apples.

In Britain the coypu appears to have no natural enemies. A bittern was recently seen to rise from a reed bed near a small bush under which were found the remains of a not-long-dead coypu; but this was in hard weather, with two inches of snow on the ground, and the animal had probably died of starvation and cold before the bittern found it. Very young coypus have been found half eaten at a marsh harrier's nest. At birth the young are about seven inches long, including the tail. They are fully covered with fur and able to swim, but do not leave the family circle until fairly well grown. Breeding may take place at any time, and each female probably rears three families a year. These average between four and eight young, though eleven is by no means uncommon.

One of the coypu's most remarkable accomplishments is that of holding its breath underwater. One which was kept in a cage before being sent to a zoo was seen to go under and remain motionless for twenty minutes, the water being just deep enough to cover it. The observer had then to go away. When he returned next day it had apparently got over its alarm, remaining on dry land to receive its breakfast of sugar-beet. Like whales and seals, coypus can restrict the flow of blood to the brain, so that the metabolism slows down. They can swim well below the surface and, when hunted by dogs, often go straight to the bottom, where they remain beyond the reach and patience of man and dog.

Despite the cold climate, coypus are bigger in this country than in Brazil. They have been known to reach a weight of twenty-six pounds in the wild state and forty pounds in captivity, although in Norfolk they average between eighteen and twenty-three pounds. In a really cold spell they cannot dive through ice and seem disinclined to burrow in the snow for roots. Their coats are apparently not then warm enough for comfort, and they sit about on the ice looking thoroughly miserable, occasionally rubbing their front paws together. Their strange cry can then be heard at all times of the day, whereas normally it is heard only at night and in the early morning. It sounds rather like that of a lost sheep, with a strange impatient note, and is audible at a considerable distance on a still night. When in an aggressive mood they also utter low growls, which can be heard only at close range.

On the Broads, where the high water table often gives rise to shallow flooding, coypus normally live in marshes covered by thick reed or sedge. Here they do not trouble to dig holes but merely make platforms to keep themselves dry. They are particularly fond of lying up during the day in a bramble thicket in or near a reed bed. When after farm crops they usually follow drainage dykes, digging large holes in the banks, where they live during the day. These holes are big enough for a small terrier to enter and turn round inside. Coypus may dig where water voles have made their holes, but by no means always do so. They have dug into a new dyke bank before any other creature has made a scratch at all. The holes are of a set pattern, entering the bank at normal water-level, so that about half is above the surface; farther in they rise just enough to allow the occupants to sit in the dry. Sometimes there is an entrance on the land side, but most of the holes are of the dead-end variety. Whole families live in them.

Although coypus are quite easily caught in open country, it is almost impossible to exterminate them in reed beds. They may change the vegetation in certain areas of thick marsh in the Broads, and other secondary effects may become apparent in time: for instance, on the birds which nest in this sort of country and may not put up with disturbance by these large and clumsy newcomers. Water rails used to breed regularly around Horsey Mere, but no nest has been found since coypus became numerous there. Perhaps the harriers will help to keep them under control, but it is unlikely that they would tackle any but the very young. Full-grown coypus are fierce when cornered and can give nasty bites with their razor-sharp teeth. There are signs that they will become a serious nuisance before long.

[This article appeared in Summer 1959 and events have proved the accuracy of the author's final forecast. – *B. C.*]

Part Nine

Various Voices

Introduction

The response to a request for eye-witness, or should it be ear-witness, accounts of frogs and toads screaming when attacked showed convincingly that this is regular behaviour; thus *The Countryman* added something to our knowledge of these familiar animals, now so scarce in many parts of the country, and showed that its readers constitute an important network of observers for this sort of inquiry.

So much has been written about bird song that I have confined myself to one contribution, a long note by Peter Driver, a zoologist who specializes in animal behaviour, but who here is lyrical in praise of an outstanding starling. Two notes by John Burton, librarian and field research assistant in the BBC's Natural History Unit and main author of the *Oxford Book of Insects*, show how modern recording expeditions can have their vicissitudes before success is achieved.

WART-BITER by John Burton

At one time the caustic fluid exuded when a bush-cricket bites was believed to cure warts; hence the scientific name *Decticus verrucivorus* of our largest and rarest kind (Illus. 22). In appearance and gait it gives the impression of a small green frog. After nearly thirty years of supposed extinction it was found again in 1955 in East Sussex and by David Ragge on damp heath close to a busy holiday road in South Dorset.

It is active only on warm sunny days, and our summer climate does not seem to be good enough for it nowadays, so few specimens had been captured until last August when I went with David Ragge to his site.

Our intention, with the help of Bob Wade as engineer, was to record the noisy stridulation of the male for the BBC's collection. The day was hot and sunny, and about six males were 'singing' – probably the most heard at one time in Britain in this century – but it took us an hour's patient stalking to catch one. The insect did not react well to captivity, and that night Bob Wade had to revive it by holding it close to an electric-light bulb. Next day, though quite lively, it remained obstinately silent; then the following morning, as the sun shone again, we heard it singing away in its box. We drove to a heath remote enough from background noises for recording, but by this time the sky was overcast and the sunloving *Decticus* had stopped singing. We had reached the middle of Poole in search of lunch when the sun came out again and a beautiful stridulation mocked us from the back of the car. Leaving the town at once, we made for the quietest place we could find in the Purbeck hills and within half an hour had recorded a full performance. We were then able to release our wart-biter on its native heath.

STARLING VIRTUOSO by P. M. Driver

Early one morning I was awakened by the sound of partridges 'jucking' on the stone tiles above my window: *skrrairp, skrrairp-skrrairp*. I stayed awake long enough to make sure that starlings were responsible and resolved to listen carefully for other calls. Since then I have heard one male give imitations of twenty-two different British birds – all so perfect in quality that I have to see him vocalizing before I can be sure it is only mimicry. My notes for a day in late October read: 'Male on roof ridge, with two birds of the year, giving mimic song "in low gear". He included magpie chatter, double crow and different single notes of cock pheasant, green woodpecker flight call, single notes of little owl, half willow warbler cadence, shrill pipe of dunnock, and *chink* and *chink-chink* of chaffinch.' I was looking through a window about twenty feet from him.

The same male apparently, in full song, began with the flight call of an immature herring gull, then gave the little owl's *keewick*, continued with a long burst of swallow song and finished with two different jackdaw calls – all mixed with his own chatter. This sequence was repeated three times in ten minutes. His imitations include flight calls of adult herring, lesser black-backed and black-headed gulls, the blackbird's alarm rattle, the single grating of the rook, the *crick* of a moorhen, the jay's screech, the female mallard's quack, the *chack* of a great spotted woodpecker and the conversational flight notes of a skein of white-fronted geese. All these birds occur regularly here on the slopes of the Cotswolds overlooking the River Severn, but no one would expect to hear all their calls within a few minutes.

The accuracy and versatility of this starling's mimicry were remarkable, but his calls have all been related in quality – timbre, pitch and general pattern – to the bird's quite varied natural song. Other observations fit this picture, such as that of the starling which imitated a telephone bell. I also remember rushing out on my lawn some years ago to investigate the sudden sound of a tribe of children playing there; they were performing by proxy, of course. Again a starling on Bardsey, off the coast of Caernarvonshire, produced quite a good imitation of an island sheepdog's barking and the whistled commands of his master.

Frog's screams

In his New Naturalist book, *The British Amphibians and Reptiles*, Dr Malcolm Smith wrote: 'During the process of being swallowed the frog sometimes utters plaintive cries, and has been heard to cry when in the gullet. Usually, however, once it has been seized, the frog resigns itself to its fate.' There is also another view: that the cry is caused simply by the expulsion, under pressure, of air from the frog's lungs over its vocal chords. In our Summer 1954 issue John Usborne described how he had heard a piercing squeal 'like that made by blowing on the edge of a blade of grass' and found a large grass snake swallowing a fully grown frog, which was 'still protesting'; the snake disgorged its victim when disturbed. Later we received from Mrs B. K. Allen a letter in which she wrote: 'Sitting by a stream one afternoon I was attracted by a noise like the creaking of a branch in the wind. As there was not even a breeze at the time, I went in the direction of the sound and was only a few yards from the water when I saw a grass snake, its head raised about a foot above the ground, looking into a hole made by a cow's hoof. At the bottom of the hole sat a frog about two inches long, looking up at the snake and squeaking.' After she had chased the snake away, she put the frog in the brook and it swam to the far bank.

As a result of these letters, a request for more eye-witness accounts was published, with an excellent response. Among the creatures which have been seen to cause frogs to scream are grass snakes, an adder and slow-worm, several hens and as many cats, a shrew, rat, mole and crow, and a number of human beings equipped with sticks. In all these incidents the behaviour of the frogs was broadly the same, although the cries struck witnesses variously as a thin high-pitched whistle, a peculiar plaintive cry and a mellow breathy scream. Always they have made a strong impression on the hearer, whether he was reminded of the scream of a young child or a rabbit's squeal; and several correspondents have expressed amazement that so small a creature could make so great a noise. One lady and her neighbour were both awakened by screams at night; on going out to investigate they found a cat playing with a frog.

The screams have usually been not only loud but prolonged. There are reports of 'a loud continuous scream' and 'cries continued without intermission'; on one occasion they continued for a full minute after a frog had been rescued from a Muscovy duck and released in an orchard. This seems to dispose convincingly of the view that the cry is caused simply by the expulsion of air from the lungs as the frog is swallowed. If further evidence is required, there are several instances of frogs screaming before they are attacked; indeed one appears to have held a cat successfully at bay by emitting an ear-piercing noise each time the cat raised its paw. A slow-worm made no attempt to approach nearer to a large frog at the other end of a doorstep, but the more it wriggled the louder were the screams. One correspondent recalls how, as a young boy, he made a frog jump repeatedly until it let out a thin high-pitched scream; and another several times saw an old farm worker wriggle a stick behind not a frog but a toad to 'mak' un cry', which it did.

Others have also mentioned toads, one of which uttered a quite prolonged cry when about to be attacked by an adder. Most remarkable, perhaps, is the incident of a large frog which was seen in a withy bed with its mouth wide open, screaming loudly, and was then grabbed by a mole; this came from under leaves and grass and, after being disturbed, returned within seconds to the attack. Some, but by no means all, observers comment that the frogs appeared to be paralysed. When a shrew leapt at the throat of a large frog sitting on a heap of leaf mould, for example, the victim uttered a piercing scream but made no attempt to escape; and the attacks and screams continued until the onlooker intervened. The habit of screaming stood many of the frogs in good stead, for they were rescued, apparently unharmed, even though they may have been halfway into the jaws of a grass snake. – *B. C.*

VOLES TAPED by John Burton

Hearing a high-pitched squealing, I looked down into a broad drainage ditch close to the River Severn in Gloucestershire and saw two water voles fighting in the limpid water. As I watched, several more appeared and I decided to make tape-recordings of their voices. So, about three o'clock on a June afternoon, we set up a parabolic reflector with its microphone across the ditch from the hole into which a vole had gone. This seemed to be the home of a pair, and both soon came into view, moving about apparently unaware of us on the roadside above. When an intruding vole swam into their territory, the male would give chase and attack it by jumping on its back with a loud splash. If the struggle was violent, the intruder uttered the loud squeal which had first caught my ear; it reminded me of the hunger cries of a nest full of well-grown young thrushes.

As we lay in the blazing sun, we also saw and recorded what I took to be courtship chases, in which the male swam after the female through the water-weeds, often jumping on her back as in his attacks on intruders; meanwhile she made a whining sound, which frequently reached a climax and became a squeal. I concluded this was courtship because otherwise the two swam peaceably together, sometimes collecting weeds which they dragged into their hole below an an overhanging willow. We also recorded at length a similar low whining sound, which I thought was produced entirely by the female or perhaps by some hidden young ones. It was above the range of the unaided ear, but over the headphones we could hear it increase in pitch and volume as the male approached its source. Not surprisingly, this sound does not seem to have been noted by the few other observers of the habits of water voles.

Part Ten

Flora's Realm

Introduction

Plants do not lend themselves to make short notes as animals do: they do not perform dramatically for *Countryman* readers on their walks abroad. But from time to time articles appear which cover different aspects of the botanist's world and four of them are offered here.

Orchids are always 'in' for interest, and Alex MacGregor's last words about chickweed wintergreen and twin-flowered linnaea almost bring the scent of Strathspey pine woods to the page. The story of the whitty pear recalls a probable ancient introduction, while David McClintock, co-author of Collins' *Pocket Guide to Wild Flowers* and author of *Companion to Flowers*, that store-house of botanical erudition, reviews some of the more ephemeral members of our alien flora, whose increase is due to the boom in budgerigars and other cage-birds. G. H. Knight concludes the selection with the sort of ecological study which makes botany exciting to other naturalists, and ends on a thoroughly contemporary note.

NORTHERN ORCHIDS by Alex MacGregor

When I asked the late A. J. Wilmott, of the British Museum Herbarium, if he had ever found a growing bog orchid he said: 'Just once, after I had slipped in a very wet bog. When I got up, I found I had sat on one.' 'Would you like to take nine flowering spikes between your forefinger and thumb?' I asked him. 'Of course,' was the reply, 'but that would be impossible.' I agreed that it would be unusual, but promised to show them to him next day. He was spending a busman's holiday at Braemar, exploring the higher glens and north-eastern corries where arctic alpines grow. These haunts were all familiar to me, as I had spent many a July day on the moors and hills of upper Deeside. On our way to a well-known glen next day I was able to show him not only the clump of nine bog orchids but some twenty others which were growing in twos and threes beside a trickle of a stream that drained several square yards of bog.

This orchid, *Hammarbya paludosa*, has a pronounced northerly distribution, occurring more widely in Scotland than in England, though its appearances are irregular even there. Owing to its tiny size and yellowish-green colour it is most difficult to see against its usual background of sopping sphagnum. Time and again in spongy parts of the moors I have been deceived into thinking I had at last located it, only to find on closer examination a few fruiting heads of the lesser clubmoss. Here, however, the flowering spikes had caught my eye as they grew erect with the base of their stems in the mosses close by the running water.

The bog orchid is one of the few which grow where the soil is acid. In place of roots it has slender hairs which are infected by a mycorrhizal fungus and ramify among mosses without entering the soil. As the leaves are small, the plant depends to a great extent on its associated fungus to supply it with food, which it stores in a swollen part of the stem just above ground. This pseudobulb serves the same purpose as a tuber in other plants, and a new one is formed each year. Fleshy knobs also develop on the tips of the leaves and

Bog orchid

later become detached. If they fall on damp moss they grow and produce new plants, more quickly than seed. The flowering head consists of a dense spike of very small flowers. Whereas in most orchids the flower stalk or ovary is twisted through a half-circle, so that the lip of the flower is brought to the front, the twist carries the blunt lip of the bog orchid through a complete circle – a phenomenon which no one has yet been able to explain.

By a curious coincidence two other orchids, coral-root (*Corallorhiza trifida*) and creeping lady's tresses (*Goodyera repens*) were first recorded for the British Isles from the

same wood in the same year (1777). John Lightfoot found them among birches near the head of Little Loch Broom in Wester Ross. I first located several spikes of coral-root among heather between the larger dunes of the Culbin sands and the sea. This was the first record of it in Morayshire, and all the more surprising because many able botanists had searched the area to list its interesting, though naturally restricted flora. In our island it is confined to the north of England and mainly the eastern half of Scotland, where it is considered rare and extremely local, though some years it may be abundant in a restricted area. It has no true root, yet it is not parasitic on living

Coral-root

plants. It absorbs nutriment from decayed vegetation by means of the fungus associated with the club-shaped portions of the root-stock which suggested its name. It is liable to die out when deprived of wet peaty surroundings by drainage. The root-stocks seem to send up aerial stems in occasional years only, so that the plant may be overlooked. But I have found it in a variety of places: among fairly tall heather, under dwarf birch, in pine woods and among pure sand overlying soft peat. In 1953 I recorded it from a new station under pines in Inverness-shire.

Full-sized spikes are six to eight inches high and carry

from four to eight yellowish flowers in a lax raceme. They may be found during the first half of June. Among heather the flowering spikes are brown as an Indian, while those in pine woods have a greenish tinge, which no doubt enables them to manufacture small amounts of food. Some years seem specially favourable for the flowering of coral-root, and I have traced groups of two to four spikes for half a mile under the outer rows of trees in an area of stunted birches. One had a root-stock almost the size of a tennis ball and had sent up five fully developed stems close together; all carried flowers at the same time.

The creeping lady's tresses orchid had three different generic names before it came to be called *Goodyera repens* in honour of John Goodyer, a Hampshire botanist who died in 1664 – a century before the plant was recorded in Britain. Characteristic of ancient pine woods, it is described in most floras as rare and local, but some recent books on Scottish wild flowers err in stating that it is rarely found beyond the valley of the Spey. A hundred years ago Professor William MacGillivray wrote in his *Natural History of Deeside* that '*Goodyera repens* is too plentiful to require the indication of stations', and in spite of timberfelling during two world wars it is still generally diffused and abundant in pine woods from Kincardineshire to Invernessshire. It even grows in heather near the sea in Morayshire, and also in two exceptional localities in Norfolk, to which it was probably introduced with soil attached to the roots of seedling pines. It is distinguished from other lady's tresses orchids by its

Creeping lady's tresses

numerous runners, which at first are covered only by sheathing scales and penetrate the thin surface layer of moss and decaying pine needles of the wood floor. In their early stages they are spotted with tufts of dark hairs which absorb food and moisture in association with the fungus infecting the stems. By their fifth year the runners develop green foliage at the tips and send down short roots, which grow stronger and thicker until supplies of food and moisture are adequate to produce flowers. When the parent stem dies, the young plants at the tips of the runners continue to grow, and in about the eighth year each carries a one-sided spike of small, sweetly scented white flowers. The spike may be slightly spiral but more often lacks the spiral twist characteristic of other lady's tresses orchids; and far from being a delicate three to four inches, as is sometimes stated, many of the erect stems reach twice that height. In pine woods in July and August this orchid is often associated with lesser twayblade, as well as with chickweed wintergreen and twin-flowered linnaea.

THE WHITTY PEAR by Augusta Paton

Three years ago a single true service tree, whitty pear or sorb (*Sorbus domestica*) was found growing on the edge of a wood in Worcestershire. It produces flowers and fruit, but, though a beautiful specimen, is by no means fully grown, being a little less than forty feet high with a girth at three and a half feet, of fifty inches. Its neighbours in the thicket are typical lime-loving trees and shrubs: ash, maple, dogwood, guelder rose and privet. Not far away grow oak, spindle, butterfly orchid, hound's-tongue, dropwort and four kinds of wild rose. Nightingales sing there in spring.

How did the tree get there? The only other known wild-growing whitty pear was the famous one in the Wyre Forest first mentioned by Edmund Pitt, an alderman of Worcester, in 1678 and maliciously burnt in 1862. Saplings from it were successfully propagated by the Woodwards of Arley Castle,

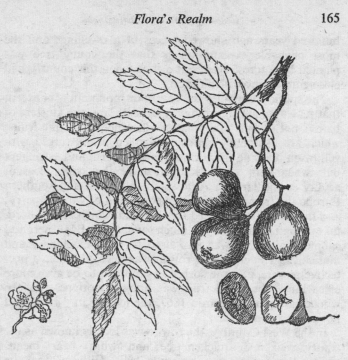

Whitty pear, after drawing by Nash

and one was planted at the original site by Mrs Woodward in 1916. Today it is about the same size as its new rival and produces fruit, though not every year. An inscription below commemorates its ancestry. Other descendants were planted at the Precentory in Worcester and, recently, near the Chapter House. One at the Oxford Botanic Garden is said to have been planted about 1790 from fruit gathered in the Wyre Forest. Two others, at Croome Court, almost certainly came from Arley stock and, though they are now no more, it seems possible that bird-sown seed from one of them gave rise to the recently discovered tree.

The true service tree is a native of the continent of Europe, and it is most unlikely that the original specimen was indigenous. An examination of its surroundings about a hundred years ago showed traces of a dwelling; and the most probable explanation is that the young tree was planted by a forest keeper at a time when this country had close connexions with France.

The species resembles the rowan or mountain ash, except that its leaves are more downy beneath and the clusters of berries not so flat. The greatest difference lies in the fruits, which are like small bunches of green pears, tinged with yellow on one side when ripe; they are harsh and astringent but, when kept until October, have the same pleasant acidity as medlars. In Anjou and in central and southern Europe the fruit was used to make a kind of wine or perry; and the tree was cultivated for its wood, which is very hard, takes a high polish and was much sought after by turners and cabinet-makers. It was also used for mill machinery and musical instruments. The rowan was deemed to have protective powers, and the sorb was supposed to be even more efficacious. The dried fruit was hung in cottages to keep witches away and, even in 1867, the idea of its virtues had not entirely died out.

In the West Country, the rowan was locally known as the whitty, wicken or quicken tree; and 'whitty pear' means simply the rowan with pear-like fruit. 'Whitty' may be derived from the Anglo-Saxon *witan*, meaning knowledge or wisdom, with reference to the tree as a charm. Another name is cheque tree, possibly from 'choker' – an allusion to the unpalatable fruit.

The original tree, described in detail and illustrated by Nash in his *History of Worcestershire* in 1781, had become a withered wreck by the middle of the last century, although the top branches still bore flowers and fruit at their extremities. It was destroyed by a poacher who had been sentenced by a local magistrate, Squire Childe of Kinlet, with a particular liking for the tree. The blackened stump and limbs were collected by a Bewdley botanist who had four goblets made from the branches; a bench of its wood was presented to the Hastings Museum in Worcester but has now disappeared, and two boxes were made by the foresters.

Winter fare

We spent the winter before last on our boat moored on Loch Dochfour, a loch which never freezes over, at the eastern end of Loch Ness. In January the smaller lochs, in which most of the winter visitants congregated, were ice-bound and the birds moved on to the open water round us. There were tufted ducks, goldeneye, wigeon, pochard, mallard and whooper swans. The only vegetation in the loch is water lobelia, which carpets the shingle bottom at depths of one to five feet and more, and this seems to provide the birds' main food supply. The base of the nearly cylindrical leaf is white and slightly thickened where it springs from the crown of the root-stock and, judging by the floating debris, this is the part the birds eat. Although enormous numbers of the leaves must be removed, the lobelia plants still carpet the bottom of the loch, so perhaps they benefit by being thinned out. – *P. R. Ivens*

The way of it

On a colliery pit-heap at Thorne, near Doncaster, a large and flourishing colony of sea campion, which is rare away from the coast, provides an interesting example of the way a plant becomes established in new surroundings. There can be little doubt that it originated from wagon-borne seed, as there are railway sidings near by. The plant has spread in a north-easterly direction from them, suggesting that the wind has been mainly responsible. A further agency now seems to be involved, for last November, a yard or two away from the nearest plants, I found collections of a dozen to two hundred calyces and capsules, all associated with evidence of rabbits, which I have seen. Partridges, too, nested on the tip last year. – *L. R. Smith*

Close harmony

On an autumn walk in Leigh Woods near Bristol I was surprised to see a heap of beech leaves under a great oak tree, because there were no beeches nearby. Then I looked at a large limb which left the trunk of the oak at right angles about ten feet from the ground. Growing out of it was a fair-sized beech; I guessed its diameter to be about ten inches. Nor was this all; beside it on the same limb was a smaller silver birch. All three seemed perfectly healthy, and the oak showed no ill-effects from acting as host to the beech and the birch. – *Lilian Devereux*

ALL FROM BIRD SEED by David McClintock

'My friend has a puzzle in her garden;' – so ran a letter I received at the end of August – 'it looks like the rare dwarf restharrow (*Ononis reclinata*), but that surely flowers much earlier.' As the plant had appeared in a place where bird seed had been scattered during the winter, it was a likely guess that it was the larger, more hairy *O. salzmanniana*, which appears in no flower book yet. This is typical of puzzlements or confusions which arise increasingly each summer, as ever more people in our affluent and kindly society feed the birds with seed mixtures. The ingredients vary from producer to producer, from purpose to purpose and from time to time, for their contents must often be affected by the availability of suitable seed. I have in mind supply and demand at the end of a hard winter, such as 1962–3, when one well-known brand seemed to consist chiefly of crumbs.

How better to identify the seeds in any mixture than by sowing them? But I suggest first spreading them out on some flat surface and grouping the different kinds. This ensures the fullest variety in the sample sown. I also retain some of each sort as a control, so that I can recognize them after I have seen the flowers. The results may well be surprising.

You will find some very ordinary plants, such as cabbage in various forms, turnip, lettuce, mustard and cress, wheat, oats, maize, sorghum, red clover, plantain and fat hen (or white goosefoot), their ordinariness not necessarily being a bar to their nutritional value. Some may be only casually included, as weeds in the crops where they were harvested. You will also get plants grown all over the world, and over the centuries, especially for their oily seeds: linseed, sunflower, safflower, hemp, Chinese or Sarepta mustard, opium poppy and nigger seed. Linseed (or flax), sunflower and opium poppy may also be grown for their beauty as annuals; so could the orange-flowered prickly safflower (*Carthamnus tinctorius*) and nigger be. (This common name comes from the black seeds; the Latin is *Guizotia*.) Hemp is sometimes grown in formal flowerbeds for its foliage, but technically its cultivation is illegal. After all, it is the source of hashish or ganja, one of the world's chief narcotics.

Linseed and hemp have been grown for thousands of years for their fibre and seeds. Linseed (like the American maize) is nowhere known wild, and so is presumably man-derived. Safflower was used to dye Egyptian mummy bands; and the history of the opium poppy also goes back to dim ages. It is fascinating to think that these plants were familiar to the Pharaohs and the Babylonians, the Ancient Greeks and the Chinese. Chinese mustard (*Brassica juncea*) looks like a hairless charlock but has prominent awl-shaped fruits. It is widely grown in the East but must often be overlooked here.

Canary grass will certainly crop up; it must grow on almost every rubbish heap in the country, as do the drooping bushy heads of the hairy millet, and the tight, cylindrical, slightly nodding heads of Italian millet, familiar in pet shops.

Among the less usuals is cow basil with large pink soapwort-like flowers; it is also sometimes sold as a garden annual. Blue woodruff (*Asperula arvensis*) resembles the too rarely grown *A. odorata* (*azurae*) but has flowers shorter than the bracts. The caterpillar plants (*Scopiurus*, or scorpion plant, in Latin and *chenillettes* in French) have small yellow pea-flowers and large, flat, often coiled, caterpillar-like fruits. The strange yellow vetchling *Lathyrus aphaca* also sometimes comes in bird seed and has straight pods for fruits. Ajowan (*Trachyspermum ammi*), an ancient spice plant, is a small white umbellifer with narrow dissected leaves. Dill is a similar ancient yellow umbellifer, much like a miniature annual fennel but with flatter fruit. The branched Indian plantain (*Plantago arenaria*) is another unusual-looking plant which sometimes occurs. *Salvia reflexa*, frequent in some years and not seen at all in others, differs basically from nearly all the others mentioned in being an American annual; it has attractive small greyish leaves and modest bluish labiate flowers. Canterbury bells and fuller's teasels (*Dipsacus sativus*) are two other possible, but biennial, surprises. The teasel, with its hooked flower bracts, was formerly much used for fulling cloth.

Then there are the puzzlers or worse, like the restharrow. In recent years the now extremely rare cornfield weed thorow-wax (*Bupleurum rotundifolium*) has been regularly reported from gardens, hen-runs and rubbish heaps. But it has always proved to be a close relative, the narrow-leaved thorow-wax (*B. lancifolium*), which has fewer stalks to the umbel, broader and more open bracts, and rough fruits. Similarly there are two species of grass poly, *Lythrum hyssopifolium* and *L. junceum*. The first is a scarce and usually fleeting annual which has been recorded in Britain for a long time and has petals only half as long as the calyx; the second, with larger flowers, appears mostly in bird seed places.

Centaurea diluta, which occurs frequently, resembles a low, spreading, pinker-rayed hardhead but has short spines on its paler bracts. *Monerma* would readily pass for a tall hard-grass (*Pholiurus*), in which genus it is sometimes put; but closer inspection would show that all but the topmost flowers have only a single glume: hence, the Latin name. *Caucalis leptophylla* sometimes appears and has been mistaken for the now rare, nearly hairless hedgehog parsley (*C. platycarpos*). The only present source for darnel seems to be bird seed. Ordinary rye-grass may well come from it as well, but darnel is a stouter stiffer annual with glumes as long as the whole floret. There is another form of rye-grass, stiff, angular, slim and often rough (*Lolium rigidum*), which may also be seen.

These are by no means all the plants which come from bird seed. Examine closely anything that appears near where such seed could have been scattered or fallen. By feeding the birds in winter, you may be rewarded with another interest in your garden in summer.

TREE WITH A FUTURE by G. H. Knight

Every countryman knows that elder flourishes at badger sets and rabbit warrens, and a few years ago I tried to find answers to two questions: what are the peculiar hazards to plants at these sites, and what adaptations enable elder to survive them? The hazards were obvious and conspicuous: both habitats are wildly unstable, suffering repeated excavations by dense local populations of mammals, which continually trample, scrape, rake, burrow and excrete and, if they are rabbits, rigorously graze all edible vegetation.

Compared with the well-endowed seeds of oak, ash, hazel and other trees, elder seeds are small and without resources. Lacking capital, they have a poor start over competitors; yet seeds planted in my garden in April produced vigorous bushy saplings of more than two feet by the autumn, whereas the other species paused at a few inches. In the wild, seeds germinate in April and average two to four inches by August; given adequate light, water and nutrients, individually favoured plants can clearly outstrip rivals. To achieve this remarkable rate of growth elder cheats. Outwardly a typical young tree, its bulk is not wood but insubstantial pith. To inflate itself to record-breaking proportions its cheap pith filling must absorb water; significantly seedlings have deep tap-roots and rarely survive outside damp shady sites. The hot dry summer of

1959 was fatal; it totally destroyed the seedlings on all the sandier sets and warrens visited. Speed is the essence of success in this shifting environment, and solidarity the price. A further result of rapid growth is that gnarled, apparently senile trees average only twenty years or less.

Elder is unpalatable to rabbits: I discovered no evidence that they eat the foliage, although they are reported to gnaw the bark when desperate. Since they graze the deadly foxglove and prickly bramble, and gnaw all other tree seedlings, this reluctance is significant. Nor does the foliage appeal to most insect larvae; in the summer of 1965 caterpillars stripped oak, birch, hazel and aspen, leaving a background of brown skeletons against the green of elder. It would be interesting to know what secret formula renders the tissues immune from grazing when better-known poisonous plants, including deadly nightshade, are devoured by certain animals.

The price of rapid growth is apparent after winter gales, when sets and warrens often become wildernesses of broken and tangled boughs; elder seems also to be defenceless against attack by fungi. The black rhizomorphs of honey agaric (*Armillaria mellea*) feel their way up under the bark or along the pith, blackening and softening the centre wood and sometimes sprouting fruit-bodies above ground. The beautiful oyster fungus (*Pleurotus ostreatus*) often joins in the attack, and many branches are flabby with the cool clammy fruit-bodies of jew's ear (*Auricularia auricula*). Despite this sinister alliance between gales and fungi, elder shows extraordinary resilience. Vigorous new branches or suckers sprout from low down. Prostrate trees root and start again; no matter how battered the victim, it usually manages a defiant sucker or two. Hedges are a permanent reminder of elder's powers of recovery; very soon after laying, elder, with ash, protrudes above the neat profile.

The tissues of elder, like those of the nettle, are relatively rich in nitrates; and both species demand nitrogen. Farmers and gardeners know that the nettle chooses the best soils. This helps to explain elder's abundance in sets and warrens, or under rookeries and starling roosts. If a plant can survive the hazards, such sites must be chemically rewarding, not only in nitrogen; and I collected some evidence that elder starts to decline in deserted rabbit warrens. It will also grow on almost pure sands, on loams and heavy clays, including the blue Lias, and this tolerance of soil types must aid its wide distribution.

The majority of seedlings and young trees exist under the shade of, for example, holly, closed-canopy oak and chestnut. These conditions discourage much summer growth but are far more detrimental to young ash, sycamore and other trees. In such habitats these seldom reach more than a few inches, whereas elder saplings of several feet are common; if there is a sudden increase in light, elder is able to exploit the situation instantly.

Elder never fails to produce a heavy fruit crop. Cow parsnip frequently flowers with it in the hedgerows, giving an opportunity to compare their popularity. Hordes of diverse insects crawl over the sticky glistening flowers of cow parsnip, but elder is ignored. It produces no nectar. The occasional hover-fly or bee lands to 'hoover' or brush off the pollen, but this is a rare event. There are reports of its pollination by small flies, but I have never witnessed this and wonder if the flies in question are the gall midges which occasionally cause the flower buds to produce 'galls' instead of flowers. Sowerby states that 'it is well known to entomologists as a favourite plant for moths', but I have not witnessed this either and find it difficult to believe that moths would be attracted to flowers which produce only pollen. I tied muslin bags round two dozen unopened umbels in mid-June, and collected these and two dozen control umbels a month later. The bagged umbels, restricted to self-pollination, averaged fifty berries; and the controls, exposed to cross-pollination, one hundred and forty-two berries. Elder's regular harvest of fruit is therefore partly explained by self-pollination, but cross-pollination increases the yield. Berries begin to ripen in August and are abundant from early September until late October. I have seen pigeons and rooks eat them, and I once watched a magpie tear off an umbel, carry it to a more convenient perch and eat a few berries before dropping it. Flocks of starlings regularly descend and gobble berry after berry; and the last warblers to be seen in autumn are often picking elegantly at the umbels.

To determine the efficiency of dispersal I searched local woods for seedlings, avoiding sets and warrens. They were not so widespread nor so abundant as, say, ash or sycamore, but sometimes they occurred in clusters below the larger forest trees. Once hundreds of seedlings were found below an aspen, far from the nearest possible parent, and they sometimes grew below birches, particularly if these were rotten and used by starlings for nesting. This puzzled me until I discovered that the birds frequented such trees and their neighbourhood well into the elderberry season. Seedlings also turned up in my town garden, invariably close to the wall or house. The birds responsible were blackbirds or, more probably, starlings again. Undoubtedly dispersal was efficient, mainly, I think, because the fruits are popular with such ubiquitous distributors.

Elder emerges from my inquiry as a typical product of contemporary life. It is light in construction, cheaply and rapidly produced, short-lived and either quickly repaired when damaged, or scrapped and replaced. The more solid durable trees like oak and yew take time to produce and are built to last; but they are hardly appropriate to the age in which we live. It may well be that elder is the tree of the future, adapted by its long association with badger sets and rabbit warrens to a countryside constantly undergoing upheaval and change.

[On the question of the elder's chemistry, which appears to render it immune from grazing, Dr S. R. J. Woodell of the Oxford Department of Botany writes: 'Elder contains an alkaloid called sambucine, a purgative resin and a glycoside called sambunigrin, which can be treated with acids to produce prussic acid. It has been used as a purgative, and its berries are diuretic. There are scattered reports of cattle being poisoned by it.']

Part Eleven

In the Water

Introduction

Life came from the water and all living things are dependent on it to different degrees: some animals spend all or part of their lives in it, others use it only for drinking and bathing. Until recently our ability to watch life underwater was very limited, but new techniques have brought the excitement of entering a new element within our grasp, as Colin Martin's article so well expresses.

But most *Countryman* readers still do their fish watching from above the surface, like E. A. C. Husbands, who saw what seems to have been a migration of the normally solitary and sedentary little bullhead. Even more curious are the bathing slugs described by Joan Blewitt Cox, a regular contributor over many years and author of two books based mainly on observations round her house and garden near Plymouth.

The rare red-necked phalarope, a wader that spends most of its time in the water, is a rarity in *The Countryman* as well and one of the few to appear in this selection. Bobby Tulloch, who writes the note to William Paton's photographs, is also the discoverer of the nesting snowy owls in Shetland, where he is the official watcher for the Royal Society for the Protection of Birds. Susan Cowdy's description of gulls gathering at Tring reservoirs shows that water birds can be spectacular in the Home Counties too. E. M. Barraud and M. Littledale observe how land birds use water, and three shorter notes are variations on this theme. H. N. Southern, who comments on the tawny owl's bath, has studied this species for years on Oxford University's Biological Estate at Wytham.

Most mammals are at home in water, like Henry Tegner's deer, though they may meet an emergency like Cecil Nurcombe's squirrel. Mink, otter and grey seal demonstrate increasing adaptation to life in the water. Andrew Neal's photographs were the first ever published of a British otter in genuinely wild surroundings and with a hitherto unrecorded prey; and it is fitting that Ronald Lockley, the senior *Countryman* writer represented in this book, should contribute the last article, about an animal which, next to the Manx shearwater, has been his special study.

FISH-WATCHING by Colin J. Martin

I became a fish-watcher in Cyprus. From afternoon swims in the warm blue sea it was a short step to an underwater mask and fins: a short step, but one I shall remember all my life. Words are not enough to describe the thrill of first coming into contact with the underwater world. It is not so much being able to see all that is going on below the surface, although that is wonderful enough; it is the sense of being part of the sea – of being accepted in a new element.

From the moment I first ducked my masked head I was an addict. I learned how to use the aqualung and was soon spending every available moment underwater. Different people have different ideas as to how to spend their time there. Some explore wrecks. Others develop an interest in archaeology, bringing pottery and implements of incredible antiquity to the surface. The more aggressive become hunters and stalk fish through the rocks with a spear-gun. I felt no desire to kill fish; they were so beautiful alive that I wanted just to watch them.

The diver has the great advantage over, say, the birdwatcher that he can move about freely without disturbing the fish. They have no fear of man, provided that he does not harm them; one shot of a spear-gun will clear an area in seconds. On a single dive it is possible to see a variety of fish behaving quite normally. As behaviour in their natural habitat has been little explored by scientists, the observant amateur can contribute greatly to knowledge in this field.

I once spent half an hour watching an encounter between a wrasse and some young bream. The wrasse, a brightly coloured little fish, was busy feeding. It would bite off a piece of coral with a snap which was quite audible underwater, and then begin to crunch it up; I could hear its munching clearly. Finally, it would spit out the odd bits of sand and dirt which it had picked up with its food, looking incongruously like somebody eating grapes and spitting out the pips.

Every now and then the wrasse would dart off and swim in a circle about ten feet in diameter round the reef. Each time it followed exactly the same path, and on one of these trips it met the bream, several of which were swimming in the area. They are lovely fish, flat and silvery with a broad black band round the tail. The adult is much bigger than a full-grown wrasse, but the wrasse I was watching met first a young one of almost exactly its own size. With a flurry of fins it chased the young bream, which nonchalantly swam off; and this happened every time a bream no bigger than itself appeared in the area. If the trespasser was larger, the wrasse took no notice of it. I could only think that the circle round which the wrasse swam on its patrols marked its feeding territory, and that intruders were not welcome. What impressed me was the unconcerned way in which the fish acted: they seemed to have a live-and-let-live philosophy, none taking the pursuit very seriously.

Then there was Horace, the octopus. I was finning gently over the sea bed when suddenly I saw an old piece of linoleum rise from the sand and place itself across an opening in the rock. This struck me as odd, so I investigated. As I drew near I got a clue from a great pile of empty shells which lay nearby; it looked like the rubbish dump of an octopus. Spellbound, I waited to see what would happen next.

After a few moments the square of lino began to move slowly down. At length a baleful yellow eye peered out: I had met Horace. He was shy at our first encounter and refused to do more than look at me from the safety of his home-made front door. The following day I returned to tempt him with chopped meat, putting some within tentacle reach of his lair. I then retired to a safe distance to watch. Before long the door opened and a sinuous arm slid out towards the meat; the tentacle coiled round it and drew it behind the rock. This was repeated twice, and each time I drew a little nearer. At length I was so close to the rock that, after I had put the meat down, I did not have to move away. When Horace came out for the meat, he saw me. 'If in doubt, give 'em ink' seems to be the octopus's motto, and very effective it is too. Everything was hidden in a thick black cloud.

I was determined to make friends with Horace and went through much the same procedure on several days. Then, just as I was expecting the usual squirt of ink, nothing happened; nothing, that is, except that Horace was slowly emerging from his den. As all I had so far seen of him was a confused jumble of tentacles, suckers and eyes, I was interested to discover his size. He was not a very big octopus; from the tip of one tentacle to that of the opposite one he measured about five feet, so each arm was only about two feet long. Now that he had broken the ice, there was no holding him. I still had some meat, and he was soon eating this out of my hand. At length he coiled himself lovingly round my neck. It was with great difficulty that I eventually got rid of him and returned to my proper element. Next time I went to see him he had gone, and his front door with him.

The so-called killers of the deep are less dangerous than popular fiction would have us believe. Certainly you could never hope to meet a creature less like a sea monster than Horace. Even sharks give no cause for undue alarm. I met a shark only once, and it was a toss up as to who got the bigger fright. On reflection I think it was the shark, which certainly showed a clean pair of fins. As skin-diving is usually carried out during the day, when a sort of truce exists between predators and prey, there is little danger of attack.

All good things come to an end; but when I had to ex-change the sunny Mediterranean for our own cold shores, to my great surprise I found opportunities for fish-watching here too. True, it is more difficult: it is colder and a diving suit is essential. Even so, the fish are just as varied in our seas as in the Mediterranean.

My real discovery came one weekend when, for lack of something better to do, I decided to dive in our local river. It was another new world – not so clear as the sea and, of course, not at all deep. Diving acquaintances look at me pityingly and say, 'The river? You'll be lucky if you get seven or eight feet of water.' But the thrill is being under the surface, whether it is two or two hundred feet above you.

I go down in this stretch of river nearly every weekend now; it has become something of a Sunday afternoon stroll. I even know some of the fish personally. There is the old salmon no one can catch, though they have been trying for years. He does not go out to sea, but just stays in his pool getting older and wiser and more battle-scarred. He does not even glance at me now, though last year he did take excep-tion when I tried to stroke him. There is a big eel, too, and I must admit to being rather scared of him. He has only one eye and moves with such ruthless purpose that I prefer to keep out of his way.

The river teems with trout and grayling. Trout are dull. Grayling are much more fun – the gossips of river life, always darting about in shoals, never still, keeping to the faster waters. They are also the most handsome of the river fish. Salmon are lordly, and trout garish; but grayling, with deli-cate striping and patterned fins, are truly beautiful. Un-fortunately they are rather shy and difficult to approach, but I found a way of getting among them. I would go upstream, dive in and let the current carry me silently down towards them. Suddenly, if I was lucky, I would burst into the middle of the shoal, catching them completely unawares. But it would be a very short glimpse, for the current would carry me inexorably on. Then came the slog upstream to get above them once again.

Strangers visit the river from time to time. Sometimes I meet a flounder, all the way from the sea, flapping lopsidedly over the river bed. They come up to feed themselves into breeding condition. Most unusual and interesting are the lampreys, which come up the river to spawn in April and May. They are among the most primitive backboned animals known, having an ancestry going back more than three hundred million years. They look pretty primitive, too, without paired fins and having a row of small openings in place of the usual covered gills. Strangest of all, they have no jaws. Instead they have a toothed sucker with a hole in the middle, and behind it a rasping tongue, which moves backwards and forwards over whatever the fish is grasping with its sucker. Unfortunately they often prey on other fish, especially salmon, attaching themselves to, and eating into, the living flesh. This parasitic habit has inevitably earned them the hatred of anglers.

Last year I arrived in the river when the lampreys were spawning. A mass urge seems to descend on a whole group of them at precisely the same moment, and they congregate to scoop out with their suckers a shallow trench in the gravel, into which they individually deposit their eggs. As with salmon, spawning exhausts the lampreys, and most of them die. The young are small worm-like creatures quite unlike the adults and take up to five years to develop.

Lampreys are supposed to make good eating, but they look so revolting that I have not dared to stomach a dish. King Henry apparently had no such scruple when he polished off the famous 'surfeit' which killed him. I did catch one to have a close look at it. It fastened its mouth on to my wrist and refused to let go until I raised it out of the water. The feel of the rasping tongue was most unpleasant.

Now I have built an underwater camera, and this summer I hope to get some really good action shots to prove to disbelievers that my fish stories are true. To any naturalist in search of adventure my advice is 'Get underwater'.

A movement of bullheads

I caught my first miller's thumb or bullhead when fishing for sticklebacks with a net, and hastily returned it to the water. At four years of age I found its big head and mouth fearsome. Seventy-seven years later, in June, I watched hundreds of these little fish moving up the River Idle at Retford. They passed me in the shallow water by the bank in twos or threes, sometimes in groups of six to a dozen, occasionally in small swarms. They swam with a flying action, settling on the bottom for a moment before setting off again; the 'flights' varied from twelve to eighteen inches. They were still coming up the river when I left a quarter of an hour later. – *E. A. C. Husbands*. [The Freshwater Biological Association has no record of a concerted movement of this kind by a fish that is normally solitary. 'Possibly', writes the Director, H. C. Gilson, 'there was an unusually dense population of bullheads in this stretch of river'. – *B. C.*]

SLUGS UNDERWATER by Joan Blewitt Cox

One evening I noticed a large orange-coloured slug mounting the pie-dish in which I put water for my tame rabbit. It leaned over the side as though about to have a drink but, when three-quarters of its body lay out across the surface, it deliberately withdrew its horns and ducked its head underwater. Then, with a lateral twist to feel for the inside of the dish, it crawled down to the bottom, remaining submerged for about half a minute but moving forward all the time. When it surfaced on the other side, its horns promptly protruded again. The slug returned to the garden, passing another of the same species, which behaved in exactly the same way. Three more followed, and later a small brown slug also submerged. Subsequently, I watched slugs bathing on many nights; sometimes none appeared, sometimes dozens collected in line, using the trails of those in front. Once I counted thirty-one coming up the garden, and there might be several together in the dish. Bathing occurred after a dry day and in pouring rain, so there seemed to be no connexion with the weather; but I wondered whether the slugs were trying to rid themselves of the mites which often infest them. What I took to be a mosquito settled on one slug's back and began bobbing up and down; through a magnifying glass I saw that it was jabbing with its proboscis. After the insect flew away, the slug stopped eating and slithered round into a U-shape so that its mouth could reach the attacked area, which it appeared to lick.

Grip on life

We think of the common limpet, *Patella vulgata*, as fixed immovably to the rock on which it is found, but a sharp sideways blow will knock one off here and there. A shallow depression exactly outlining the shell is then revealed. This air-and-watertight seal between shell and rock helps it to withstand the roughest seas. Limpets feed mostly on small growths of weed on intertidal rocks and must go in search of them. A night visitor to a thickly populated stretch of rocks hears 'a rustling that seemed like a bustling', sees by the light of his torch the limpets moving about gently but persistently, and can pick them up quite easily. If a limpet and its stance are marked by day, a search the next night the rock is uncovered will show how far it has travelled. The oldest and biggest tend to go farthest, perhaps nine or ten feet. Each must return to its stance after feeding; if it is not home when the tide covers it and the sea is rough, it will lose its grip on the rock, which to a limpet is its grip on life. – *Donald McGregor*

Ladies only!

The islet on which the photographs were taken (Illus. 32) is only an acre or two in size, with one little pool. It is used almost exclusively by female red-necked phalarope as a gathering and feeding-place, while the males are occupied with domestic duties on the main island a mile or so away; for in this species the normal roles are reversed. On the rare occasion when a male appears on the islet, he often seems to spark off an excited burst of chasing during which female chases female, female chases male or male chases female. This may end in a flurry on the water, and a female will flutter momentarily on the back of another, or, as in the lower photograph, on a male. I have seen this happen only once or twice, and it was not followed by mating. The gathering in the photographs was the most prolonged, with the most persistent chasing, I have yet seen; I think eight birds were involved, and at times all the females were flying after the male. – *Bobby Tulloch*

COMING IN TO ROOST by Susan Cowdy

It is awe-inspiring, on a frosty winter afternoon, to see the white hordes of black-headed and common gulls glide in on angled wings against a red-and-orange sunset sky to roost on the Tring reservoirs, which now form one of their winter inland bases. Once they are down the chatter and muttering begins, and by four o'clock a close-packed flock is noisily splashing and preening on the water. Overhead, parties of up to a hundred are still flying in, some in formation so that they appear from a distance as skeins of geese, often giving rise to wild goose stories. Their method of landing varies. Sometimes they make a direct approach, planing down to join the rest without fuss; but they usually spiral over the water for some minutes, often to a depth of a hundred or more feet, before they settle.

R. A. O. Hickling, who organized a winter gull investigation for the British Trust for Ornithology, has suggested that the spiralling may assist the many new arrivals to join without congestion the countless birds already on the water. When at last the late-comers decide to land, they do so neatly and surely. At this time the gulls are extremely nervous. If they

hear an unaccustomed noise, such as a distant gunshot, there will be a sudden silence followed by a mighty whoosh of wings, as they all leave the water and start to spiral upwards in massed flight. They drift away on the wind out of sight, and the reservoir seems strangely still and empty, apart from a raft of duck over by the far reed-beds. Beyond, the Chilterns range darkly against the twilight sky and parties of rooks fly from them to roost in trees round Mentmore, a few miles to the north. Then you become aware again of thousands of silent specks overhead. The gulls are returning. They spiral down but are this time reluctant to land, until one braver than the rest drops lightly on to the water. The others quickly follow and the chatter breaks out once more in the great concourse, now barely visible in the gathered dusk.

BIRDS BATHING by E. M. Barraud

One of my earliest bird-watching discoveries was that there are few more rewarding sites than the bank of a stream. Many

times I have gone miles to such a spot. Only recently did I
realize that I had in my garden an excellent potential lido:
a spring-fed ditch at the far end of the orchard. In December
I cut right down a rather scruffy elm hedge; and during the
spring I kept the herbage trimmed as it began to shoot. At
one end of the twenty-yard stretch of newly opened water I
built a wigwam hut with a moveable front, where I could sit
in a deckchair within a few yards of the drinking and bathing
birds. My 'estate tally' of birds seen in the garden totals some
seventy species, and I saw nineteen of them at the water in
the first week of watching: no rarities but such (for us) less
usual birds as the turtle dove, pied wagtail, tree sparrow,
moorhen and coal tit. It is not rarity that fascinates me
anyway, but the ordinary things ordinary birds do in the
course of their ordinary lives.

A first I cursed the ubiquitous house sparrows, always
present by the dozen. But almost the first thing I discovered
was that there seemed to be no competition or aggression
over the water. Birds do sometimes squabble at a bowl, but
in this natural drinking-place even 'individual distance' was
apparently ignored. Species and individuals mixed freely,
bathing and drinking close together without friction. I saw
a pair of goldfinches, a pair of greenfinches and a robin all
within a few feet of each other, and a family of bullfinches
mixed with a young starling, a cock blackbird and a hedge
sparrow in complete amity. Once I had realized the astonishing absence of aggression, I saw that the sparrows were more
of an asset than a liability; several were always in the water
and their splashing attracted the passer-by, who at once
came down for a drink and a bathe too.

I soon noticed that most of my visitors bathed twice.
After the first relatively brief dip the bird would fly up to a
tree, often not preening at all, and at most giving only
cursory attention to its plumage. Then it would come down
again for a more prolonged and complete wetting. Only
after that did it fly up to a branch for a thorough preen. At
first I noticed this only with the less common birds; but then
I saw it even in sparrows, after Baldy (a female with no

feathers on her nape) gave me means of individual identification. I now believe that it is the rule rather than the exception for birds to take two baths, with an interval of between three and ten minutes, unless they are disturbed.

Next I was surprised to see how many birds indulge in what the ethologist calls 'intention movements' before bathing. Time and again a bone-dry bird, high on the bank or even perched on a tree, shivered its wings, puffed out its feathers and tossed its head, just as if it were in the water: obviously 'thinking about it', like a small boy standing on the river bank. I had often seen these intention movements in caged birds, when a companion was occupying the water bowl; but I had no idea that the habit was common among wild birds, even when quite alone near the water.

I had long known that birds had pretty accurate internal clocks, being seen day after day at the same times and places; but this was particularly striking at the bathing-place. The goldfinch pair were punctual to within ten minutes, even when they started to bring with them their four grey-pate young. Some species pairs were closely companionable, among them linnets and goldfinches; but the chaffinches

never came together or even
in any regular relation to
each other's visits. Of the
bullfinches, the female
arrived first (and later
brought her young) and the
male always about ten minutes later.

The mechanics of drinking also varied. Starlings slid their
open bills under the water before raising their heads to swallow, whereas most other birds quickly dipped their beaks and
raised them. The pigeons, of course, drank steadily, like
cattle. Robins, blackbirds and thrushes hopped and jumped
about, facing all ways, before settling down to wet themselves; tits, robins and sparrows squatted low in the water
and really soaked their plumage, whereas linnets and goldfinches stood 'tall' on their legs, flicking the water up over
their bodies from this standing position. All the birds I
watched bathing used their wings alternately, as well as both
together, to throw water over their backs. In some species
this was very pronounced, and I was reminded that I had
often seen tame pigeons do it in slow motion, actually
pausing when they were heeled right over in the water.

I had planned to continue my regular watch through the
summer and into the winter; but one hot day I found only a

waste of mud in the bed of
the stream. I guessed what
had happened. There was
now a good flow in the long-
dry ditch in front of the
cottage, the farmer up the
road having diverted the
stream to make a pool of
water to cool his churns. Not
till the end of October was
their rain enough to supply
both channels, and by that
time bathing would have

dropped off greatly, even if the birds had not lost the habit of going to my ditch.

From November onwards there was no sign of goldfinches, greenfinches or linnets, which is not surprising, since great flocks of them were then to be seen over the fields in open country. Nor were there any sparrows; they had moved into the rickyard after harvest and had been bathing and drinking at my stone sinks. But I did see at the ditch the orchard blackbirds (and a few strangers), robins and thrushes, some tree sparrows, small flocks of great tits and the odd woodpigeon. Bathing was now a much quicker affair – often quite perfunctory – although I still saw 'double' bathing and the alternate wing movements.

Now I am looking forward to the warm days of another spring, and the hot days of summer. Meanwhile, I shall have time to carry out one or two minor clearings and improvements before my customers come back.

Magpie's swim

On a warm April day I was sitting in the garden not far from an iron water-tank about four feet deep and filled to the brim, when a magpie settled on the edge of the tank. It ducked its head in and out of the water; then, to my amazement, it jumped right in and had a good splash before it settled on the rim again. This performance was repeated three times. I concluded that there must be a lid inside the tank, a few inches below the surface of the water; but there was nothing of the kind. How was the magpie able to keep afloat and rise from the water with such ease? – *Hazel Inglis*. [Nearly all birds can swim for a short time, but this magpie's behaviour was certainly unusual. – *B. C.*]

Dew for two

In a slight depression of the lawn the longer grass was holding an extra load of dew, in which a dunnock was fluttering, streaking through the blades in ecstatic enjoyment of its dew bath. A blue tit was edging closer and closer to the fanatical bather, until its approach aroused active resentment and it was forced to retire for a few moments. At last the dunnock stopped its antics, ruffled itself, preened one or two feathers and flew off. No sooner had it gone than the blue tit returned and started copying its evolutions, like a tiny speed-boat fluttering spray in all directions. – *A. T. Hitch*

OWL'S BATH by M. Littledale

The illustrations (Nos. 33–6) show a young tawny owl from a deserted brood hand-reared on chicken giblets. When fully grown they returned to the wild and, though still appearing to regard humans as friendly, would not accept food or allow a close approach. The whole bathing incident took about half an hour. At first the owl was puzzled at seeing water and slipped into it from the cement bank. After a startled squeak, it dipped its head deeply and threw water over itself with evident pleasure. But it looked woebegone when it fluttered out and tried to go on washing itself, banging its beak on the ground and again squeaking with surprise. Then it discovered it could dry off with a dust bath, which it much enjoyed. Still damp and dusty, it fluttered on to a sunny patch of dry grass, where it walked and preened until it was dry enough to fly off. [The passion of tawny owls for bathing is notorious, though we know far less about its application to other species of owl. Wherever water-tanks are left uncovered in fields and woods, sooner or later a tawny owl is sure to be found drowned there. They seem to have difficulty in judging the depth of water in artificial sites. Such deaths are not due to inexperience, because I have usually found the birds killed in this way to be adults. The present account suggests that the tawny owl learned to bathe by accident. It would be interesting to know whether an activity so obviously enjoyed always needs such a chance stimulus to start it off. – *H. N. Southern*]

AQUATIC DEER by Henry Tegner

My observations suggest that nearly all kinds of deer are at
home in water; in fact, many seem to enjoy an occasional
dip. There have been recorded instances of Scottish red
deer swimming the seas to and fro between the mainland and
some of the Western Isles; and when an island becomes over-
populated, starvation or evacuation are the only alternatives.
Several years ago Sika deer became so numerous on Brown-
sea Island in Poole Harbour that some swam to the Dorset
mainland, where they quickly established themselves.

The River Tyne, as it approaches Newcastle, is a wide
polluted stretch of filthy black water; yet quite recently a
young roe buck was found swimming a mile upstream from
the city. There are many roe in Northumberland and Dur-
ham, and they cross freely the upper reaches of the Tyne; but
the choice of a densely populated stretch of the river seems
strange. The animal had probably been disturbed, like a roe
buck I watched in the summer of 1960; frightened by a collie,
it entered the Garry when the river was in spate, made a
wide detour in the birch woods on the far bank, then returned to the river and swam across it again under a steep
waterfall. A sea-going roe buck was rescued not long ago off
the coast of Normandy, where a party of fishermen saw it
swimming in the Channel and took it on board.

In the Scottish Highlands red deer will ford or swim
across any normal stretches of water they encounter. Rivers
like the Dee, Don, Tay and Spey are but trickles to them. In
the west of England red deer regularly cross the Exe and the
Barle. Fallow deer, too, will enter water when travelling from
one feeding ground to another. I have often seen them do it
in Northumberland where there are regular deer fords in the
Aln and other rivers. Deer will also enter streams and lakes
to feed on the underwater vegetation. In an Angus deer
forest during a period of deep snow a red deer hind discovered
succulent feeding on the bottom of an unfrozen lochan. She
soon had numerous others of her kind savouring the lush
underwater growth.

Assisted Passage

We had visited an island of about five acres, a quarter of a mile offshore in the Oslo Fjord, and were returning to Larkollen because of the freshening wind. I was rowing from the centre seat of the boat. About halfway across I noticed a red squirrel swimming desperately after us in the now turbulent water. Its tail was straight out behind it and partially immersed at times, perhaps due to the choppiness. I backwatered until it came close, then held one oar steady, thinking it would clamber on to the blade; but it turned towards my wife, who was sitting with our daughter in the stern. She extended an arm, placing her hand in the water. Without hesitation the squirrel climbed on to it, ran up the arm and jumped from her shoulder to the gunwale. Apparently distressed to find two people in the stern, it ran forward past me and my son to the prow, where it rested until we came within yards of the landing-stage. Then it plunged into the water, swam ashore and continued its journey without looking back. – *Cecil Nurcombe*

Mink at large

Certain rivers in Devon are becoming infested with escaped mink, which are now breeding freely. My first sight of one was when salmon fishing with a neighbour, who suddenly shouted: 'Look over there! See the mink?' Quite aware of us, it frisked and frolicked along the opposite bank. When it came to an inlet, it swam across, shook itself and went on its way. It was very dark brown, larger than I had expected – almost as big as a cat, but nearer the ground and in behaviour a little like an otter. I saw my second mink when I was fishing alone with my springer spaniel Saxon for company. As I was unravelling a 'bird's nest', I heard a plop; a mink emerged from the river immediately in front of me, shook the water from its coat and advanced towards me. I do not think it saw me. Saxon's ears were at the alert, and he looked as amazed as I was. The mink came right up to the toe of my thigh-boot and paused, seemingly surprised by so unusual an obstacle. Saxon began a low deep growling and his hackles rose. I prodded the mink with the butt of my rod. It was simply staggered but, recovering itself instantly, did a beautiful dive back into the river, leaving on the surface a line of bubbles. – *Penelope Cook*

OTTERS AT PLAY by T. R. Barnes

One day towards the end of August I was sitting in a punt munching my sandwiches and trying, unsuccessfully, to 'trot' my float through the weed-beds. On one bank were water-meadows; on the other tangled and decaying willows grew out of a patch of impassable bog. Earlier in the morning, while fishing downstream, I had heard in this area a curious high-pitched yakking chatter which, in my ignorance, I attributed to 'some sort of water bird'. Now I heard the sound once more and saw a movement among the weeds that fringed the bog. A full-grown otter slipped into the water and swam towards me. It then turned and called again, and another slightly smaller otter swam out. Both came within ten yards of the punt, but neither took the slightest notice of me. They proceeded to play, rolling over and over, diving and chasing round and round. They would surface together, come face to face with bared teeth, put paws each on the other's shoulders and duck each other, then bob up a few yards away and go through the game again. This was varied by complicated chasings and rollings, but the face-to-face ducking was the favourite manoeuvre; and all the while I was completely ignored. At first, of course, I froze; but later I continued eating my sandwiches, as it did not seem to matter if I moved. The game went on opposite the punt for about ten minutes; then both animals, still playing and 'talking' vigorously, gradually drifted down with the current. I could see nearly a quarter of a mile downstream, and an angler on the bank about three hundred yards below me stood up when the otters came opposite him, but still they took no notice. Nearly half an hour must have elapsed after the start of their play before they went out of sight and hearing.

OTTER AND DOGFISH by Andrew G. Neal

Last summer I was exploring a tidal island in Loch Sunart, which stretches about twenty miles along the southern shore of the Ardnamurchan peninsula in Argyll, when I saw an otter swimming towards the weed-clad rocks. It dived suddenly and reappeared only ten yards offshore with a fish in its mouth. Keeping out of sight, I crawled forward with my camera. For a moment I thought the otter must have seen me and swum away; then I spotted it beautifully camouflaged against the brown seaweed at the water's edge. It was so intent on its prey that I managed to approach within thirty feet and take several photographs before it saw me and dived into the water, popping up a few seconds later from behind a rock to have a last picture taken (Illus. 37–9).

The prey, a nineteen-inch spotted dogfish, had not been eaten at all, only opened at the belly, although the otter had been wrestling with it for two minutes or more. It might have made a meal of its catch if I had not disturbed it. Alternatively it may have killed it by mistake, not realizing it to be such a tough-skinned species, or for sport. Otters reach adult length (forty to fifty inches) when fourteen to eighteen months old. I judged this one to be between thirty-five and thirty-eight inches, and so perhaps ten to twelve months old. At that age it should have been a fairly experienced judge of its prey.

THE BLIND LADY OF LITTLE SKERRY
by R. M. Lockley

The traveller by air between Scotland and Kirkwall, the cathedral village-city of the Isles of Orkney, looks down on two specks of land awash in the snowy wool of tides roaring through the Pentland Firth between the North Sea and the Atlantic Ocean. The larger speck, Muckle Skerry, is recognizable from its cluster of neat lighthouse buildings, as white as the racing tides. A mile to the south, Little Skerry is a dark salt-drenched slab like a giant black whale with head ploughed to the east, striving in the creaming current.

It was alongside a rocky flipper to the north of the Little Skerry whale, after days of waiting for the rare right weather and a certain quiet condition of the neap tide, that the crofter's lobster-boat sidled for a few seconds, barely long enough for us to jump ashore: 'Haff an 'oor! Haff an 'oor! Ye canna bide mor'n a wee while the day.'

The day was full of calm wintry sunshine. Marvellous to be on this lonely skerry so seldom visited by man; to finish for a while with the wild jumping of the saucer-shaped, if seaworthy, Stroma boat (built almost as broad as long to counteract the confused movement of the Pentland sea); to be on dry land in the company of five hundred Atlantic grey seals, more than a hundred of them mothers nursing their calves.

Hardly dry land: the few acres above high water were strewn with the remains of wrecked ships – anchor, wheelhouse, taffrail, waterways – and the patches of soil and the rocks were wet with salt and slimy with the slug-like journeyings of the adult seals and their passionate squirming and twisting in territorial and libidinous activities. All around they lay, staring at us in amazement and alarm, hesitating to escape until we had shown our intentions. Their way to the sea was over the jagged north-side ledges, dangerous for an animal who cannot do more on land than hunch along on fore-flippers and stomach. We, too, needed to move warily; the rocks were extremely slippery, and we did not wish to disturb, only to count, the seals, putting down on paper the age composition of the herd. The less we moved about the better. We had to crawl to the top of the island, seal-wise, to prevent a panic; seals are short-sighted but recognize man as an enemy of upright carriage.

The island plateau was parcelled out with family parties like Hampstead Heath on Bank Holiday. Each massive adult bull dominated a harem of a dozen cows and their calves. The biggest bull sat regally atop the crown, close to the wheelhouse of a wrecked trawler which rested drunkenly with gaping sides. In the doorless doorway a mother seal was suckling a fat yellow baby. In round figures (we muttered, trying to get in a quick count) there were one hundred and ten calves, one hundred and twenty-eight mothers, twelve bulls (first-class or beachmasters), six secondary bulls and, say, two hundred and fifty virgins, young bulls, adolescents and young hangers-on – an easy five hundred, including those in the sea. In our excitement we stood up to get a view all round. Instantly there was a general exit of adults, who slid and slipped through a brown pudding carpet of soil and vegetation, of bird and seal droppings, towards the sandstone ledges and the sea.

The king seal lay staring at us balefully while the dirtied white calves scattered and bleated at the departure of his cows. He would not move but opened his huge mouth in a leonine snarl to illustrate what our fate would be if we approached closer. We did not. He breathed slowly, heavily, with a distinct hiss between his scarlet tongue and white canine teeth, nicely hooked and sharp for tearing fish, Near the master bull there sat, unmoving, one lone dirt-encrusted cow, her flipper protectively over her once-white calf. Head erect, she looked the picture of proud motherhood, unafraid of man; but closer inspection showed that she was quite blind. She could hear and smell us, but she did not move (Illus. 40).

Arctic and antarctic seals, born on the sterile ice-fields, do not suffer infections like these grey seal calves dropped into the muck of soil, faeces, rock debris and decaying seaweed. As they move or are pushed around by the adults, the skerry calves become covered with septic scratches, and many have their eyes stuck up with dirt and pus. Probably this blind cow had caught an eye infection when a calf, perhaps on this same insanitary skerry, and had gradually lost her sight over the years. Nevertheless, she was in good condition, as fat as any other seal matron of many seasons.

As we gazed at those filmy eyes and then came near, almost to touch her, we thought of her problems in getting to her present nursery niche on the highest, safest spot on Little Skerry. Had she delivered her calf here each year before she became blind? And afterwards, had she returned each autumn, feeling her way by touch slowly over the ledges whose topography she would remember from past use? Obviously she could manage to catch fish underwater, so why should she not be able to move on land? As for the problem of parturition, it is now suspected, though difficult to prove, that when the cow seal reaches her full term, she can delay delivery of the calf for several hours until conditions of tide, place and weather are suitable. Mercifully, this blindness of our Little Skerry lady allowed her to be uneasy but not to panic, as the other cows had done, on our arrival. Her unease at the noise and scent of man beside her was weaker than her maternal instinct; so long as she could feel and smell her baby – as illustrated by her petting it with one fore-flipper and sniffing it at intervals – she was reassured.

The old blind lady, the fishermen told us on the sail homewards, had been known to them for many seasons, and they had tried to follow her life at sea as well as during her annual appearances on Little Skerry. They believed she was able to get a good living browsing the sea floor with her sensitive whiskers and swallowing squid. Lobstermen of Orkney protected seals, not superstitiously because of their legendary connexions with Viking ancestors, but more practically because the 'selkies' took squid and octopus which preyed on the valuable shellfish. Wherever they set their pots among the skerries, they got a better haul as long as the seals were about; so good luck to all seals, blind or seeing. Moreover, the lobsters liked the seals and followed them around, picking up the titbits from their feasts of fish.

Index to Plants and Animals

Adder:
  and cats, 24–7
  takes willow warbler, 63
Ants, taken by swallow, 69

Badger:
  caught in gate, 150
  nests above ground, 118–21
  and tame vixen, 56
Bats:
  in hair, 28–9
  take moths, 65
Bats, Horseshoe, 29
Bats, Pipistrelle, 28
Beech growing on oak, 168
Bees:
  food of hedgehog, 62
  food of toad, 61
Beetles, food of shrew, 54
Birch growing on oak, 168
Birdseed, plants in, 168–71
Blackbird:
  bathing, 185
  image-fighting, 104
  and little owl, 83–4
  and rat, 86
  and sparrow, 72
  and squirrel, 86
  tame, 51
Bream, and wrasse, 177

Bullfinch:
  bathing, 185, 187
  feeding, 39–40, 42
Bullhead, movement of, 181
Bunting, Red, on Skomer, 24
Bush-cricket, 154–5
Butterfly:
  Bath white, 136
  large white, 137–9
  Painted Lady, 137
  Peacock, in house, 139–40
  Purple Emperor, 140
Buzzard:
  back-hacking of, 46–50
  and crows, 89
  from pony, 30
  and weasel, 89

Campion, Sea, inland, 167
Cats:
  and frogs, 158
  and little owl, 88
  and magpie, 88
  and shrew, 88
  and snakes, 24–7
Chaffinch:
  bathing, 186
  feeding skills, 76–7
Chat, Robin, image-fighting, 105

Chickadee, taken by shrike, 71
Chiffchaff, on Skomer, 23
Chortophila spreta, 60–61
Corallorhiza trifida, 162
Cormorant, and gull, 82
Coypu, 150–3
Crossbill, in felled tree, 32–3
Crow:
  and buzzard, 89
  feeding skill, 77–8
  image-fighting, 104–5
  and little owl, 83–4
Crow, Carrion:
  aerial play, 100–102
  play on roof, 106
Crow, Hooded, and rabbit, 87
Cuckoo:
  lays in box, 130–1
  reared by flycatchers, 129–30
  reared by meadow pipits, 129
Curlew:
  and fox, 74
  robbed by starlings, 73
Curlew, Stone, broods in rain, 127–8

Decticus verrucivorus, 154–5
Deer, Fallow:
  aquatic, 190
  habits, 143–7
Deer, Red, aquatic, 190
Deer, Roe:
  aquatic, 190
  watching, 18–21, 29–30, 31
Deer, Sika, 190
Dipper, builds nest, 121
Dog:
  beagle and fox, 97
  corgi and hares, 91
  labrador and fox, 147–8
  and squirrel, 91

Dogfish, and otter, 193
Donkey, 'groomed' by jack-daws, 122
Dormouse, mistakes man for tree, 34–5
Ducks:
  on ice, 68
  on Skomer, 21–2
Dunnock (hedge sparrow), bathing, 185, 189

Earthworm chased by mole, 64
Eel, jackdaw and gull, 73
Elder, ecology of, 171–4
Emperor, Purple, 140
Epichloe typhina, 60

Fieldfare, tame, 43–5
Flies:
  in filter beds, 59
  fungus-eating, 60–61
Flies, Crane-, taken by fox, 65
Flycatcher, Spotted, rears cuckoo, 129–30
Fox:
  and beagles, 97
  eats crane-flies, 65
  and labrador, 147–8
  and mallard, 74–5
  tame vixen, 54–6
Frog:
  screams, 156–8
  taken by stoat, 37
Frog, Marsh, and snake, 37

Gnats, taken by birds, 69
Goldcrest display, 111
Goldeneye on Skomer, 22
Goldfinch, bathing, 185, 186
Goodyera repens, 162, 163–4
Grayling, and skin-diver, 179

Greenfinch:
  bathing, 185
  feeding skills, 76–7
Gull, Black-backed:
  and cormorant, 82
  on Skomer, 23
  roost at Tring, 183–4
  takes starling, 72
Gull, Common, roost at Tring, 183–4
Gull, Herring:
  and eel, 73
  play with leaf, 103
  scaring off, 109–10
  on Skomer, 23

Hammarbya paludosa, 161–2
Hare:
  and dog, 91
  and stoat, 92–3
Harrier, Marsh, and coypu, 151, 153
Hedgehog:
  circular walk, 36
  eats bees, 62–3
  parasites on, 149–50
  self-anointing, 36
Hens, plucked by starling, 122
Heron:
  and fox, 74
  on Skomer, 22
  tameness, 42–3

Jackdaw:
  and gull and eel, 73
  pecking order, 66–7
  takes donkey's hair, 122
  tame, 51–2
Jay:
  anting posture, 106
  and weasel and mouse, 97

Kestrel:
  grips wire, 111
  and rat, 96
  and stoat, 94, 96, 97
  and vole and weasel, 97

Lady, Painted, 137
Lampreys, and skin-diver, 180
Lapwing, in snow, 127
Limpet, on rock, 182
Lobelia, Water, food of waterfowl, 167

Magpie:
  and cat, 88
  image-fighting, 105
  swimming, 188
Mallard:
  descent of young, 123–4
  and fox, 74–5
  on Skomer, 21
Martin, on Skomer, 24
Martin, House:
  in hand, 52–3
  sunbathing, 107
Mink, in Devon, 191–2
Mole:
  breeds in cage, 117–18
  chases worm, 64
  takes frog, 158
Moorhen:
  builds nest, 124
  on Skomer, 22
Mosquito, taken by spider, 58, 59
Moths:
  camouflage of, 140–2
  Clifden Nonpareil, 141
  Common Shark, 141
  Peppered, 141–2
  taken by bats, 65

Mouse:
 and jay and weasel, 97
 and kestrel and stoat, 96
Mouse, Long-tailed Field or Wood:
 joins picnic, 148
 rescues young, 124–5
 and robin, 88

Newt, sloughs skin, 35
Nuthatch, feeding habits, 40–41

Oak, host to beech and birch, 168
Octopus, and skin-diver, 177–8
Orchid:
 Bog, 160–2
 Coral-root, 162–3
 Creeping Lady's Tresses, 162, 163–4
Otter:
 and dogfish, 193
 at play, 192–3
Owl, in flight, 19
Owl, little:
 and cat, 88
 takes thrush, 82–4
Owl, Tawny, bathing, 189
Oystercatcher, and starlings, 73

Partridge, by cooking-fire, 32
Peacock Butterfly in house, 139–40
Peregrine, aerial play, 102–3
Phalarope, Grey, on Skomer, 22
Phalarope, Red-necked, in pool, 182–3
Pheasants, territorial behaviour, 84–6
Pigeon:
 'buzzes' car, 108
 plucked by sparrow, 122
 sunbathing, 107
Pintail, and fox, 74
Pipit, Meadow:
 rears two cuckoos, 129
 on Skomer, 24
Pipit, Rock, on Skomer, 24
Polypedilum lobiferum, taken by birds, 69
Pony, as observation post, 29–31
Pteromalus puparum satellite, 135

Rabbit:
 and elder, 172–3
 and hooded crow, 87
 and sea campion, 167
 and stoat, 90–91, 93–4
 and woodpigeon, 87
Rat:
 and blackbird, 86
 and kestrel and stoat, 96
Rat, Brown:
 eats snails, 63–4
 and stoat, 94–5
 takes frog, 37
Raven:
 aerial play, 99–100
 play on roof, 106
Redstart, Black, 34
Redwing in flight, 19
Robin:
 bathing, 185, 187
 hand-tame, 45
 and mouse, 87–8
 on Skomer, 23
Rook:
 aerial play, 102
 feeding skill, 78
 and rabbit and stoat, 96–7
Rowan, 166

Ruff:
  on Skomer, 22
  take gnats, 69

Salmon, and skin-diver, 179
Sandpiper:
  on Skomer, 22
  tameness, 43
Seal, Grey, blind female, 193–6
Shark, and skin-diver, 178
Shieldbug, maternal care, 115
Shoveler, on Skomer, 22
Shrew, Common, tame, 53–4
Shrew, Pygmy, and cat, 88
Shrike, Bokmakierie, image-
    fighting, 105
Shrike, Great Grey, in Essex,
    70–71
Shrike, Puffback, image-fight-
    ing, 106
Shrike, Red-backed, nest, 128
Skylark, taken by weasel, 92
Slow-worm 'duel', 81
Slugs:
  food of shrew, 54
  underwater, 181–2
Snails, food of brown rat, 63–
    4
Snake, Grass:
  takes frogs, 37, 157
  takes toad, 81
Snakes and cats, 24–7
Sparrow, Hedge (dunnock),
    bathing, 185, 189
Sparrow, House:
  bathing, 185–6
  and blackbird, 72
  image-fighting, 105
  intelligence, 78–9
  malformed beak, 70
  plucks pigeon, 122

  and song thrush, 72
  and sparrowhawk, 71
Sparrow, Rufous, image-fight-
    ing, 105
Sparrowhawk:
  from pony, 30
  on roof, 71
Spider, 'Money', 59–60
Spider, Spitting, 58–9
Squirrel, Grey:
  and blackbird, 86
  and dog and wasps, 91
  feeding skill, 75–6
  and stoat, 90
Squirrel, Red:
  builds drey, 116
  swimming, 191
Starling:
  bathing, 187
  before roosting, 110–11
  feeding, 66
  as mimic, 155–6
  play on roof, 106
  plucks hen, 122
  robs curlew, 73
  sunbathing, 107
  taken by gull, 72
Stint, on Skomer, 22
Stoat:
  at bird table, 75
  and hare, 92–3
  hunting methods, 92–5
  and kestrel, 94, 96, 97
  and mouse, 96
  and rabbit, 90–91, 93–4
  and rat, 94–5, 96
  and squirrel, 90
  takes frog, 37
  winter coat, 92, 94
Stonechat, on Skomer, 24
Sun birds, image-fighting, 106

Swallow:
  feeding habits, 68–9
  revived, 32
  on Skomer, 24
Swallowtail at Wicken Fen, 132–6
Swan, rescue of young, 126–7

Teal:
  and fox, 74–5
  on Skomer, 22
Thrush, Kurrichane, image-fighting, 105–6
Thrush, Mistle:
  attacks fieldfare, 44, 45
  hand-tame, 50–51
Thrush, Song:
  and little owl, 82–4
  multiple nests, 128
  and sparrow, 72
Ticks, on hedgehog, 149
Tit, Bearded, taken by shrike, 70–71
Tit, Blue:
  feeding, 68
  two nests, 129
Tit, Great:
  feeding, 41–2, 50
  sunbathing, 107
Tit, Long-tailed, removes feathers from nest, 123
Tit, Marsh, feeding, 39
Toad:
  and bees, 61–2
  and grass snake, 81–2
  scream, 158
  and wasps, 62
Trout, and skin-diver, 179

Vole, from pony, 30

Vole, Field, and weasel and kestrel, 97
Vole, Short-tailed, hand-reared, 53
Vole, Water:
  courtship chase, 159
  voice of, 158–9

Wagtail, image-fighting, 105
Wagtail, Grey:
  on Skomer, 24
  young rescued, 31–2
Wagtail, Pied:
  image-fighting, 103–4
  on Skomer, 24
Warbler, Willow:
  on Skomer, 23
  taken by adder, 63
Wasps:
  food of toad, 62
  remove wood, 116
  and squirrel, 91
Weasel:
  and buzzard, 89
  and mouse and jay, 97
  rescues young, 125–6
  and skylark, 92
  and vole and kestrel, 97
Weaver, Golden, image-fighting, 105
Whitty Pear, 164–6
Wigeon, on Skomer, 22
Woodcock display, 112–13
Woodpecker:
  in felled tree, 33
  in snow, 67
Woodpigeon:
  in felled tree, 33
  and rabbit, 87
Wrasse, and bream, 177
Wrens, in blackbird's nest, 66

Index of Contributors

Adams, M. C., 34
Alexander, Lolita, 50–51
Allen, Mrs B. K., 157
Alsop, S. E., 62
Anderson, E., 89–90
Archer, W. H., 105

Barnard, Gwendolen, 147–8
Barnes, T. R., 192–3
Barraud, E. M., 184–8
Batterbee, H., 72
Beswick, J. S., 50
Bevan, Alan S., 88
Blakiston, Ann, 125–6
Bodman, Frank, 121
Bodman, Janet, 111
Bosworth, Arnold, 67–8
Boul, W. T. G., 65
Brade, P. G., 108
Brown, R. W., 62
Burnett, Brian W., 92, 104–5
Burton, John, 154–5, 158–9
Buxton, J. J., 150–3

Campbell, A. M. G., 140
Campbell, Stewart, 87
Carne, P. H., 143–7
Carnegie, Sacha, 18–21
Chapple, Fred J., 124–5
Cohen, Betty, 104
Cook, Penelope, 191–2

Cooke, C. H., 107
Cooke, J. A. L., 59–60
Cooper, Anne L., 124
Cowan, James Ker, 88
Cowdy, Susan, 183–4
Cox, Joan Blewitt, 181–2
Cranbrook, Earl of, 27–9
Cullin, Basil, 65

Dalton, Stephen, 58–9
Davidson, S. R., 75
Dearnley, J., 91
Devereux, Lilian, 168
Douglas, Elizabeth W., 86–7
Driver, P. M., 155–6
Durand, A. L., 34–5

Edwards, Frank C., 78

Faithfull, E., 84–6
Fishwick, J. W., 82–4
Fussey, Joyce, 66

Gait, R. P., 43, 128
Gardiner, Brian, 132–6
Gittins, J. W., 76
Glasier, Phillip, 127–8
Goodwin, Derek, 104, 108
Gordon, J. P., 52–3

Hacker, W., 51–2
Hale, P., 96–7

Hamilton, George H., 109
Hamilton-Price, Keith, 45
Hardiman, Joyce, 73
Harper, S. Winifred, 64–5
Harries, Sheila, 66–7
Harrison, R., 122
Hart, Mary, 31–2
Hawkins, Desmond, 96
Hemingway, J. E., 96
Hewson, Raymond, 92–5
Hilton, A. C., 115–16
Hitch, A. T., 188–9
Holland, Marjorie, 126
Hopper, Florence, 91
Howard, Len, 39–42
Humber, R. D., 32–3
Hurrell, H. G., 45–50
Hurt, Freda, 70
Husbands, E. A. C., 181

Inglis, Hazel, 188
Ingram, Molly, 105
Ivens, P. R., 167

Jacobs, C. J., 78
Jennings, R. J., 97
Jones, H. Stanley, 122

Kabell, Margaret B., 63–4
Kear, Janet, 76–7
Kelcey, Nancy M., 68
Knight, G. H., 171–4
Knight, Maxwell, 35–7
Knowlton, D., 90–91

Langman, Alan, 148
Lennon, P. J., 129
Letts, Vivienne, 54–6
Littledale, M., 189
Lockley, R. M., 193–6
Löhrl, Hans, 106

Lovenbury, G. A., 90
Lumley, Ida, 139

McCaffrey, John, 42–3
McClintock, David, 168–71
MacGregor, Alex, 160–4
McGregor, Donald, 182
McHattie, D., 89
McLean, J. T., 87
Manini, Emily, 69
Mann, Stella V., 86
Marshall, S. C., 97
Martin, Colin J., 176–80
Mears, Chris, 106
Middleton, Elizabeth, 110–11
Miller, F., 35
Molloy, H. A., 72
Montrose, Anne, 71–2
Mylne, C. K., 148–50

Neal, Andrew G., 193
Neal, Ernest, 118–21
Newman, L. Hugh, 136–9
Nurcombe, Cecil, 191

Oliver, G. R., 86

Palmer, Ray, 115
Paton, Augusta, 164–6
Pearce, M. E., 68–9
Pearson, Andrew, 53
Peterson, John, 82
Pettitt, Florence E., 130–1
Phillips, A. B., 72
Phillips, Greta D., 106
Pitches, F., 78–9
Platt, John E., 111
Price, Bernard G., 32

Reade, Winwood, 67, 91
Reeman, Frank G., 117

Renateau, L. P., 33
Richards, T. J., 99–103
Ricketts, Violet I., 81
Rickman, Mary, 105
Ripley, M., 122
Robbins, Helen, 123
Rudkin, Heather K., 68

Saunders, D. R., 21–4
Shepherd, R. T., 66
Sherdley, Charles, 123
Shimmer, Kay, 116
Shorten, Monica, 116
Smith, L. R., 167
Snape, Winifred, 32
Sole, Molly, 104
Spearing, Dorothy, 64
Stevenson, F. C. W., 77–8

Tarn, W. H., 62
Taylor, Hilda N., 150
Taylor, J. S., 69
Taylor, Mark, 73–5

Tegner, Henry, 29–31, 112–13, 190
Theak, Doris, 129–30
Thompson, R. M., 72–3
Tinbergen, Niko, 109–10
Tinklin, Richard, 75
Tudge, J. A., 81–2
Tullis, J. H., 61
Tulloch, Bobby, 127, 182–3
Turk, Stella M., 53–4
Tutt, H. R., 70–71
Tweedie, M. W. F., 140–2
Tyler, Stephanie J., 63

Usborne, John, 157

Veall, Patricia, 69
Ventress, H. O., 43–5

Watson, Thomas, 61
Wells, G. E., 24–7
Williams, C. B., 60–61
Wright, Doris, 62

THE MOST SOUGHT AFTER SERIES IN THE 70's

These superb David & Charles titles are now available in PAN for connoisseurs, enthusiasts, tourists and everyone looking for a deeper appreciation of Britain than can be found in routine guide books.

BRITISH STEAM SINCE 1900   W. A. Turpin   45p
An engrossing review of British locomotive development – 'Intensely readable' – COUNTRY LIFE. Illustrated.

LNER STEAM   O. S. Nock   50p
A masterly account with superb photographs showing every aspect of steam locomotive design and operation on the LNER.

THE SAILOR'S WORLD   T. A. Hampton   35p
A guide to ships, harbours and customs of the sea. 'Will be of immense value' – PORT OF LONDON AUTHORITY. Illustrated.

OLD DEVON   W. G. Hoskins   45p
'As perfect an account of the social, agricultural and industrial grassroots as one could hope to find' – THE FIELD. Illustrated.

INTRODUCTION TO INN SIGNS
Eric R. Delderfield   35p
This beautifully illustrated and fascinating guide will delight everyone who loves the British pub.   Illustrated.

THE CANAL AGE   Charles Hadfield   50p
A delightful look at the waterways of Britain, Europe and North America from 1760 to 1850. Illustrated.

BUYING ANTIQUES   A. W. Coysh and J. King   45p
An invaluable guide to buying antiques for pleasure or profit. 'Packed with useful information' – QUEEN MAGAZINE. Illustrated.

RAILWAY ADVENTURE   L. T. C. Rolt   35p
The remarkable story of the Talyllyn Railway from inception to the days when a band of local enthusiasts took over its running. Illustrated.

GAVIN MAXWELL

RAVEN SEEK THY BROTHER (illus.) 30p

Again set in the lovely West Highland surroundings of Camusfeàrna, this is a remarkable self-portrait full of marvellous photographs, anecdotes, descriptions of people and landscapes, birds and animals, and times of comedy and tragedy.

'Splendid for lovers of animals and fine writing' – THE EVENING NEWS.

LORDS OF THE ATLAS (illus.) 40p

The breathtaking story of the meteoric rise and spectacular fall of a warrior Berber tribe, and their almost legendary leader – T'hami El Glaoui, Pasha of Marrakesh.

'Romantic and horrifying' – THE TIMES.

RING OF BRIGHT WATER (illus.) 30p

Now an enchanting film starring Bill Travers and Virginia McKenna.
Camusfeàrna, a remote home in the Scottish Highlands, a pet otter as intelligent and affectionate as a dog, this is both a moving and a fascinating account of the author's homelife.

THE ROCKS REMAIN (illus.) 30p

This is the beautifully-written sequel to RING OF BRIGHT WATER which describes, among other things, the arrival of the new otters Teko, Mossy and Monday.

'As beautifully written, as vivid, and as moving as its predecessor' – THE GUARDIAN.

A SELECTION OF POPULAR READING IN PAN

FICTION

SILENCE ON MONTE SOLE Jack Olsen	35p
COLONEL SUN	
A new James Bond novel by Robert Markham	25p
THE LOOKING-GLASS WAR John le Carré	25p
A SMALL TOWN IN GERMANY John le Carré	30p
CATHERINE AND A TIME FOR LOVE	
Juliette Benzoni	35p
THE ASCENT OF D13 Andrew Garve	25p
THE FAR SANDS Andrew Garve	25p
AIRPORT Arthur Hailey	37½p
THE GOVERNOR'S LADY Norman Collins	40p
SYLVESTER Georgette Heyer	30p
ROSEMARY'S BABY Ira Levin	25p
A KISS BEFORE DYING Ira Levin	25p
HEIR TO FALCONHURST Lance Horner	40p
THE MURDER IN THE TOWER Jean Plaidy	30p
GAY LORD ROBERT Jean Plaidy	30p
THE THISTLE AND THE ROSE Jean Plaidy	30p
TRUSTEE FROM THE TOOLROOM Nevil Shute	25p
THE VIRGIN SOLDIERS Leslie Thomas	25p

NON-FICTION

THE MONEY GAME 'Adam Smith'	35p
THE SOMERSET AND DORSET RAILWAY (illus.)	
Robin Atthill	35p
THE WEST HIGHLAND RAILWAY (illus.)	
John Thomas	35p
RAVEN SEEK THY BROTHER (illus.)	
Gavin Maxwell	30p
MY BEAVER COLONY (illus.) Lars Wilsson	25p
THE PETER PRINCIPLE	
Dr. Laurence J. Peter and Raymond Hull	30p
THE ROOTS OF HEALTH Leon Petulengro	20p

These and other advertised PAN Books are obtainable from all booksellers and newsagents. If you have any difficulty please send purchase price plus 5p postage to P.O. Box 11, Falmouth, Cornwall.

While every effort is made to keep prices low, it is sometimes necessary to increase prices at short notice. PAN Books reserve the right to show new retail prices on covers which may differ from those previously advertised in the text or elsewhere.